ARTICULATED AGRICULTURAL DEVELOPMENT

Articulated Agricultural Development

Traditional and Capitalist Agricultures in Papua New Guinea

MIKE DONALDSON

Department of Sociology
University of Wollongong

and

KENNETH GOOD

Department of Political Studies
University of Zambia

Avebury

Aldershot · Brookfield USA · Hong Kong · Singapore · Sydney

Published by
Avebury,
Gower Publishing Company Limited,
Gower House, Croft Road, Aldershot, Hants GU11 3HR,
England

Gower Publishing Company,
Old Post Road, Brookfield, Vermont 05036,
United States of America.

ISBN 0 566 05469 8

Printed and bound in Great Britain by
Athanaeum Press Limited, Newcastle-upon-Tyne

Contents

Preface

The following study examines a process of vital importance in the economic and social development of many Third World countries. This is the procedure through which so-called traditional or non-capitalist societies become interlinked or articulated with new capitalist agricultural systems. The articulation of the old and the new socio-economies is, however, gravely problematical. The parlous conditions of agriculture today in many Third World countries, and hence of their economies generally, is testament that the successful transition to commercial agriculture is far from being the obvious, pre-ordained, or common outcome. For this articulation to be carried through, many important factors must be present in both the non-capitalist and the capitalist modes and must together play their part in conditions where conflict is assuredly not absent, but where cooperation and mutual support also occurs between key institutions and social groups of the different systems. It is a process of uncertainty and complexity even under favourable circumstances. But if a symbiosis is achieved between the non-capitalist agricultural society and the new economy, a viable commercial agriculture solidly established in the hands of peasants, that is of large numbers of indigenous producers, may well result. This is not a negligible outcome. The development of an efficient agriculture is of supreme importance to the strength and well-being of a country and it can serve as the essential basis for the further and integrated development of manufacturing industry. It is equally important in colonial and decolonising situations that indigenous people rather than foreign interests should acquire control over domestic resources, and the engagement of peasantries in commercial agricultural production, rather than the involvement of a national bourgeoisie in manufacturing, is the feasible development in many places.

Articulated Agricultural Development is the analysis of the successful articulation of the old and the new agricultural systems in the highlands of Papua New Guinea (PNG). Here is a case of the implantation in less than thirty years of agricultural capitalism, centring around coffee production for export, among experienced and receptive non-capitalist farmers. While aspects of articulation have been studied in other places, the existing works have tended to concentrate on the theoretical level, on the non-capitalist stage as such, and upon the limitations or failure of the articulation. [1] The transition to commercial agriculture in PNG has, by contrast, been relatively smooth, quick and extensive, and this study examines the different and changing factors contributing to this notable outcome.

That this was so has been due in no small part to the strengths and forms of non-capitalist agriculture in PNG. The articulation became possible because of the direction and degree of development of the existing social formations, particularly of the dynamic and hierarchical features within them. These tendencies represented a particular and strong receptivity for commercial agriculture when this began to be available through the agency of Australian colonialism beginning in the 1950s. Among the major factors contributing to the transition to peasant based cash crop production were those considered earlier by Amin with regard to Africa: the old societies should be sufficiently hierarchical to permit certain individuals to appropriate land, and also labour, at an early stage; the population densities should be neither too high nor too low for land and labour to become individually controlled; good and suitable cash crops must be available; and the state must at least not be hostile to the development of peasant based commercial agriculture. [2] Other studies of dissimilar situations have stressed the importance of the surplus producing capabilities of the earlier systems and the availability of new commercial markets for cash crops. [3] These and other specific features were at work within the differing modes in PNG, but it was the whole ensemble of interacting factors which made possible the agricultural transition there. Particular features of the new capitalist mode were insufficient in and of themselves, neither the policies of the colonial and post colonial state, nor the opportunities of new markets, were of independent determining importance to the articulation.

This study is, however, not one of unalloyed admiration over a significant developmental outcome. The breadth and detail of the study permits an accurate and balanced assessment of the trajectory of extensive rich cash crop production which has been undertaken by the peasants in PNG. The costs and benefits of this transition are considered at the societal level as they directly involve and affect the majority of coffee growers. They are also examined at the individual level of those particular rising rich peasants and rural capitalists who have done best out of the growth of commercial agriculture. The costs include the increasing inequalities and antagonisms between the great majority and small minority of farmers. But the assessment recognises both the pre-existence of inequalities in PNG, and their integral and originating position in capitalist development everywhere. The actually existing developmental alternatives in PNG as in most other Third World countries, were and are extremely narrow, and nothing which could be portrayed as 'socialist' agriculture was either growing out of the past or is on the agenda for the future. The other available item on the developmental programme of the colonial and post-colonial state was actually minerals' exploitation by multinationals.

Notes

[1] The work of the French marxist anthropologists Meillassoux, Rey and Terray, has focussed on approaches to pre-capitalist social formations. See for example David Seddon (ed.), Relations of Production: Marxist Approaches to Economic Anthropology, London, Frank Cass, 1978; for the theoretical concerns see Harold Wolpe, 'Introduction', in Wolpe (ed.), The Articulation of Modes of Production, London, Routledge and Kegan Paul, 1980; and for the failed articulation see Peter Geschiere, 'The Articulation of Different Modes of Production: Old and New Inequalities in Maka Villages, South-east Cameroon', African Perspectives, 2, 1978.
A recent collection of essays on modes of production in Africa calls for an increased concentration on studies of the non-capitalist period; Donald Crummey and C.C Stewart, 'The Poverty of Precolonial African Historiography' in Crummey and Stewart (eds), Modes of Production in Africa: The Precolonial Era, Beverley Hills and London, Sage Publications, 1981, p. 12.

[2] Samir Amin, Accumulation on a World Scale: A Critique of the Theory of Underdevelopment, New York and London, Monthly Review Press, 1974, pp. 366-367. The above listing extrapolates slightly from Amin's original formulation.

[3] Colin Bundy, 'The Emergence and Decline of a South African Peasantry', African Affairs, 71, 285, and Ian Phimister 'Peasant Production and Underdevelopment in Southern Rhodesia, 1890-1914, with Particular Reference to the Victoria District', in Robin Palmer and Neil Parsons (eds), The Roots of Rural Poverty in Central and Southern Africa, London, Heinemann, 1977. Both South Africa and Rhodesia were settler states strongly opposed to the development of a peasant based commercial agriculture.

1 Agricultural production and class formation

Agricultural production in Latin America, Asia, Africa and the Pacific is at once of obvious and primary importance. [1] It is usually the basis of a country's economic and social reality whether expressed as the leading sector in gross national production, in the value of exports, or in the way of life of the great majority of the people, or in all three. Many countries are typified by the production of particular commodities: the once and future 'King Cocoa' of Ghana, cloves of Zanzibar, sugar in Cuba, coffee in Kenya and, of course, the ubiquitous Banana Republic. In Papua New Guinea (PNG) coffee was the leading agricultural commodity at the end of the 1970s. Coffee, cocoa, and copra - the so called 'Three C's' - represented some 50 per cent of total export earnings, and coffee alone was worth approximately 60 per cent of this figure, and was, moreover, that sector wherein indigenous ownership was by far the most advanced.

When the grain and root crops of many places are accounted for, the saliency of agriculture is such that three-quarters of world population depend on the products of their fields and agricultural labour for their livelihoods and agriculture represents two-thirds of the export earnings of the Third World. More than 70 per cent of the labour force still works the land in Africa, producing between 30 per cent and 60 per cent of the gross domestic product (GDP) in most places. [2] Only minerals and petro-chemical production, for example copper in Zambia, oil in Nigeria, to a lesser extent copper in PNG too, rival agriculture in export earnings but not in people's livelihood. Manufacturing is important, especially as part of diversified, integrated and relatively advanced economies, chiefly in the past and present settler states. In strength or in weakness, agriculture is the fabric of most Third World countries in the broadest sense.

The actual recognition of the saliency of agricultural production is, however, partial at best. The colonial powers were certainly conscious of the agricultural resources of the Third World, and a central objective of colonialism has been to obtain food and other commodities at the lowest possible price. In the post World War Two conjuncture of limited reserves of arable land, rising populations, food shortages and of Cold War and decolonisation, United States' agencies and corporate interests have been inclined to view agriculture as a lever for control in the Third World. The United States Agency for International Development (USAID) was fairly explicit in its recognition:

> ... virtually all countries we assist are primarily rural with 70 or more percent of the people earning their living in agriculture. US assistance to the agricultural sector of the developing countries is therefore, to an exceptional degree, significant in the influence it has in the ideological contest ... the fundamental political character of the country which emerges from the development process will be determined largely by the rural institutions designed to achieve agricultural development. [3]

Other agencies were aware of the importance of the United States' position as 'custodians of the bulk of the world's exportable grain', and the strategic opportunities which this offered to 'retain the primacy in world affairs'. [4]

European colonial interests in general were most effectively represented in a colony which was approaching independence, Smith has suggested, by 'an [agricultural] import-export elite whose capacity to develop economically ... allow[ed] it to cooperate usefully with the international system and at the same time assure domestic stability'. [5] The extension of many 'food aid' programmes by the United States and the former colonial powers grew out of such assessments and were utilised, dependent upon time and place, to further the United States' international political and economic goals. The 'Food for Peace' programme in India in 1965-66 for example, was administered so as to gain advantages for United States' corporate interests and to control the local fertiliser industry, [6] and early in 1981 food aid to a then drought stricken Mozambique was abruptly terminated by the United States following the expulsion of two American diplomats. The United States has often used agriculture as a weapon against Third World countries of which it disapproves. It closed its important domestic market to the sugar exports of Nicaragua under its Sandanista government, and it has assisted rebel forces concentrating in 1984 upon the destruction of Nicaragua's vital coffee and cotton export production. [7]

Scholars recognised the criticality and breadth of agricultural production rather more slowly and narrowly. Only in the 1970s was agriculture being seen more clearly as both a changing way of life and a dynamic system of production with domestic and international linkages. An interdisciplinary, historical and materialist approach was heralded by works such as Palmer and Parsons' <u>The Roots of Rural Poverty in Central Southern Africa</u>, and Iliffe's consideration of the 'origins of rural capitalism' and 'the creation of tribes' in modern Tanganyika. In 1980 a history of agriculture in PNG also appeared. [8]

The concern with the origins of rural change necessarily embraces anthropological perspectives in addition to those of history and economics. Anthropology had a long and leading position in the social science about the non-Western world and its literature constituted a valuable repository of data on domestic economies, food resources, and colonial situations. [9] The data was, however, embedded in a theoretical bias which emphasised the importance of distribution and the notion of kinship. [10] Social anthropology under the influence of Malinowski and Radcliffe-Brown also stressed the scientific and practical policy aspects of the discipline and was inclined to promote the study of phenomena such as so-called cultural contact and assimilation in association with colonial governments. [11] Almost a decade and a half before the new interdisciplinary and historical study of agriculture, however, a new marxist anthropology appeared which focussed on production in pre-capitalist societies. [12] The perspective here was expressed by Coquery-Vidrovitch, 'Exchange, the most obvious manifestation of the network of relationships, reflects the internal organisation of society, but it is the result of the organisation of production rather than its cause' (emphasis of the original); and by Terray who said following Marx that it was 'in production that we must look for the secret of the social edifice'. [13]

One of the generalised contributions of the new anthropology was its partial identification of certain myths concerning non-capitalist agricultural systems, that they were subsistence, homogeneous and largely static. Logically, agricultural surpluses must have been experienced at least sporadically within a region if specialisations such as salt, pottery, tool or weapon making were to be supported by communities therein. [14] In addition to the exploitation of slave and serf labour in certain social formations in Africa and PNG, [15] appropriation and exploitation existed in non-capitalist agricultural social formations in relations such as those between male elders and both young men and women, and between chiefs and aristocrats and most other members of a social group.

The development of agriculture, with its inter-linked, cyclical activities and deferred outcomes, perhaps inevitably involved the appearance of privileges of seniority. Young men and newcomers became dependent upon elders for nourishment, and access to essential resources such as land, tools, seeds, and knowledge. The latter not only tended to prolong this situation advantageous to themselves but frequently their domination included and was defined by control over women's labour and its product. [16]

Deepening exploitation and the gradual emergence of classes were part and parcel of the development of non-capitalist agricultural production. [17] The appearance of surpluses, appropriation, and inequalities accelerated the process of social change. [18] Non-capitalist agricultural formations were neither timeless nor static, although Terray suggested that change was slower where domination was based directly on the 'natural' factors of age and sex. [19] There was, for example, both a wealthy class and poorer people at Great Zimbabwe, there was a clear distinction between the court and ordinary people in the Mutapa state, and in the later development of the Shona pre-colonial formation women, children, slaves and bondsmen could be more clearly identified as the weak and powerless. [20] The exploitation of women's labour was a notable feature of the development of non-capitalist agricultural production in the highlands of PNG.

The identification of developing inequalities and classes over time in pre-capitalist systems suggests the appropriateness of seeing class formation as being a slow, interrupted and uneven process. The study of the formation of classes must involve both a consideration of the structural relationships to the means of production, and the historical analysis of the process of formation and action within which classes are fashioned by the changing forces of production and simultaneously shape them. Commonly, because of the juxtaposition of different modes of production within a given formation - dependent on time and place involving perhaps hunting and gathering, non-capitalist and capitalist agriculture, and trade, class appears as an elusive and complex phenomenon. [21] It must also be recognised that non-property based structures of association and conflict such as ethnicity, race, religion and language 'interact with cleavages derived from the economic sphere to prevent a simple crystalisation of class conflict along property lines'. [22] But class is also an active and originating developmental force, a point which Brenner has emphasised in the wide historical sense, [23] and which was borne out in Africa and PNG in, say, the thrusting, acquisitive action of chiefs, elders and big men, and in the sometimes more equivocal and latent opposition of the weak and poor.

Despite such realities, PNG studies have been peculiarly blind to the importance of social change. The weight of the anthropological stereotype of the 'timeless present' and the bias against production has lain heavy on the land. There has been a fixed approach to the 'clan' and social organisation, and a minimisation of the fluidity of these institutions in the past and present. Changing patterns of land ownership in areas of advanced agricultural production have often been ignored, and the significant question of the relationship between so-called 'tribal' fighting and class action has sometimes been denied out of hand. [24] Thus, the not atypical statement: 'The reader's imagination is stretched to the limit ... when told that fighting in the Highlands [today] is the result of the spread of capitalism and the emergence of the rich peasantry as an exploitative class'. [25] The strong possibility that the development of rural capitalism has a growing effect on groups, ownership and disputes assuredly deserves attention without denying the fact of intense warfare in the pre-colonial highlands or asserting the over determining effects of contemporary capitalism.

Class interprets social change in general, but it also plays a big role in the articulation of different modes, most critically of the capitalist with non-capitalist modes. Rey referred to the 'operative value' of the term class in that it could 'explain the present in terms of alliances between the dominant classes of two separate and articulated modes of production ... and that it would be extremely difficult to explain this in other terms'. An indigenous exploiting class is always a 'useful ally' for a foreign bourgeoisie because the former is 'embedded in production itself and it provides an immediately available base for the transfer of surplus labour to capitalism'. [26] But a foreign bourgeoisie is sometimes a useful ally for the leading indigenous classes too, and the intrusion of agricultural capitalism, especially in the form of rich cash crop production, can represent big and portentious opportunities for the advance of already powerful social entities. The rise of 'King Cocoa' in Ghana is properly understood only when it is seen in relation to an enthusiastic domestic acceptance and incorporation on a differentiated social basis. [27] Class formation and action are probably the leading and most dynamic forces at work within the articulation.

Class in agricultural production under the increasing sway of capitalism and the state focusses essentially on formation and action by peasantries. In Shanin's well known analytical definition, the peasantry 'consists of small agricultural producers who, with the help of simple equipment and the labour of their families, produce mainly for their own consumption and for the fulfilment of obligations to the holders of political and economic power.' [28] This definition has the advantage of stressing the inherent dependency of the peasantry upon and market, but it ignores the large variations which often exist among sections of the peasantry with regard to the usage of equipment and labour. Peasants are still 'the majority of [hu]mankind', [29] and are usually the domestic producers of food and of cash-crops in the Third World. The established field of peasant studies is therefore important as another source of data and of comparative perspectives on the later trajectory of non-capitalist agricultural production and the rise of rural capitalism. [30]

Peasant studies are however, almost as rich in myth and controversy as anthropology. Some Eurocentric marxists even refuse to recognise the existence of 'peasantry' as a theoretical category, and simply subsume the peasant majority as part of a huge, undelimited petty bourgeoisie. [31] Approaches to the peasantry are generally characterised by a dichotomous view, according to both Worsley and Lieberson. For the former, there exists a distinctive 'inward' and an 'outward' view of peasantries, [32] while for the latter the 'moral economy' approach is sharply contrasted with the view of the 'rational peasant'. [33] The 'inward' view is associated with writers in the Chayonovist school and with substantivist anthropologists and, says Worsley, stressed the supposedly unique character of peasant culture and the supposed autonomy of the internal workings of the peasant household and village. This attitude is similar to 'the moral economy' approach which Lieberson sees as 'something of an orthodoxy', and as represented by the work of James C. Scott and Clifford Geertz. The peasantry are here portrayed as 'noble savages' and their societies as characterised by 'collective solidarity', a 'subsistence ethic' and a distinctive 'moral universe' (in Scott's terminology). Goran Hyden's recent work on Tanzania and on Africa generally is further indicative of the orthodox position. He too emphasises the uniqueness of the African peasantry and recognises an homogeneous peasant society and a 'moral economy' or 'economy of affection'. The so-called 'peasant mode of production' is presented as largely static and as rather sharply and totally opposed to capitalism. Hyden is therefore inclined to portray rising commercial farmers as no longer peasants and tends to minimise the contemporary importance of richer peasants within agricultural development. [34]

By contrast, Lieberson's minority position, represented for him by the work of Samuel L. Popkin, identifies the inequalities and conflict present in pre-colonial peasant societies, and sees at least certain elements of the peasantry as potentially entrepreneurial and readily disposed to innovate if personal gain is involved. This is essentially similar to the 'outward' view, where the household and village are not autonomous but a segment of a wider economy and society wherein the non-peasant sector dominates the peasant part.

The full truth about peasant societies possibly lies between the dichotomous positions but much closer to the 'outward' perspective which alone has a capacity to identify the hierarchical, dynamic, and potentially developmental tendencies within agricultural societies. The 'inward' or orthodox view serves to disguise the changing position and role of rising peasant classes within the development of agricultural production in the Third World generally and in the highlands of PNG particularly. The 'outward' approach to the study of peasantries is, therefore, relevant to an understanding of the process and period of the initial articulation, its continuation, and aftermath. Within Third World social formations peasantries are the main domestic classes of the articulation with capitalism. The direction of pre-capitalist and commercial agricultural development in the highlands of PNG appears to approximate what has been termed the Kulak path: [35] the internal differentiation of the peasantry proceeds in such a way as to enrich a small minority of peasants and impoverish others. [36] And this small class of rich peasants, in contrast with the view of Scott, Geertz, and Hyden, may be seen to have emerged out of and retained important linkages with the dynamic and hierarchical elements of the earlier and still partially extant non-capitalist agricultural modes.

The analysis of the changing international environment in which the domestic economies of Third World agricultural production are placed is the particular concern of dependency or underdevelopment theory. The early work of Frank and of Griffin appeared shortly after that of the new marxist anthropology, [37] but little dialogue existed between the schools for some time. The weaknesses in this early work have been much commented upon. [38] There was a strong deterministic tendency which visualised the necessary perpetuation of dependency in the Third World, and in its earlier formulations there was a neglect of class as an active and original force. However, the gaps and minimizations in the original dependency notion do not negate its central insights. [39] It focussed in quite an original way upon the global expansion of European capitalism and it stressed the 'crucial importance' of this process, as Roxborough correctly notes, to Third World social formations. [40] The corrollary was that it was valueless (though all too common) to study such societies in isolation from the conditioning of the world system and from their location in history. The dependency approach saw the impact of global capitalism representing the greatest force for change in most Third World countries, while recognising (in its mature forms) that development therein has not been a mere reflection of or response to exogenous factors.

Dependency might therefore be understood, following Roxborough, as a way in which the complex articulation of several modes of production within a given Third World social formation is achieved. [41] This articulation is to a significant degree derivative of the expansion of the world economy, and works in ways that either encourage only certain forms of domestic development, such as the over-reliance upon weak and unintegrated commodity producing systems, or which transfer the benefits of growth abroad, or both. But less totalistic and more differentiated outcomes should also be considered. Rich cash crop production in the hands of peasantries, for example, would seem to represent a different and lesser form of dependency than, say, a monoeconomy of foreign owned copra, sisal, or banana plantations, or minerals production by multinationals.

The emphasis on the mode of production, perhaps the special contribution of marxist anthropology, facilitated a more systematic and detailed investigation of domestic change. Mode of production analysis attempts to identify with some precision key social totalities, their constituent parts and changing inter-relationships. O'Laughlin said that the concept is used 'both to define conceptually a particular dialectical (and therefore dynamic) unity of forces and relations of production and, at a different level of abstraction, to delimit a period of history dominated by a particular mode of production'. [42] The term social formation is utilised partly in order to avoid deceptive suggestions of homogeneity and cohesiveness often conveyed by the term society. It also indicates the complex unity developed within the separate individual modes of production, and between the levels of the different modes in the articulation. In its entirety, this complex, dynamic, unstable, yet unitary phenomenon, is the social formation. [43]

Indeed, the complex articulation of different modes is, as noted, a characterising feature of most Third World social formations in the contemporary period. 'Determination' within and between these modes is to be understood as a 'setting of limits' or an 'exerting of pressures' rather than as 'control'. [44] The actual linking together of different modes is partly expressed through alliances of leading social classes and entities in the different systems. The domination of one mode over another exists when the functioning of one is subject to the requirements for the reproduction or expansion of the other, [45] an earlier hunter-gathering mode becoming subject to a mode based on food and fodder production, for example, and later both being made subject to the development needs of metropolitan capitalism. Analysis therefore focusses upon key shifts or transformations in production and social organisation within and between social formations. Relations of production, modes, and social formations are bound together with class and exploitation, and the use of one such concept presumes the use of the others. [46]

The key problem for much of the Third World is the development of viable, integrated commercial agriculture based upon the successful articulation of non-capitalist and capitalist agricultural systems. The analysis of this problem obviously requires a broad conceptualisation and a dynamic, interdisciplinary approach. Class formation considers the active element in development and the key agencies in the transition, wherein rising peasantries are the salient domestic social entities. Class and peasant studies offer strong correctives to the dangers of abstraction and scholasticism which have appeared in marxist anthropology, [47] but the capacity for precision in mode of production analysis is important. As O'Laughlin concluded, 'The constructs mode of production and social formation provide no automatic discovery procedures: they do specify the direction of analysis and help us to organise what we know'. [48] What we wish to identify, Anderson has said, is 'a set of lines of force for transformation ... [which] in turn incarnate certain values that are an active part of the process of social transformation itself'. [49] Articulation represents a potentially critical transformation in the Third World, and the development of agricultural production, class formation, and the action of the state embody forces and values which contribute powerfully to its success or failure.

Notes

[1] This chapter is derivative of Good, 'Relevant Theory: Production and Class Formation in Their Societal-Historical Settings', Australian Outlook, 36, 1, 1982.

[2] Rene Dumont and Nicholas Cohen, The Growth of Hunger, London and Boston, Marion Boyars, 1980, pp. 5 and 44; the Economist, 10 October 1981, p. 90; and for PNG see, e.g., 'Copper Swings and Coffee Roundabouts', Economist, 30 September 1978. A key aspect of Zambia's weakness, for example, is that agriculture represents less than 14 percent of GDP.

[3] Statement by USAID officials in January 1964 and quoted in Susan George, How the Other Half Dies, Harmondsworth, Penguin, 1977, p. 73.

[4] A document of the Central Intelligence Agency published before the 1974 World Food Conference and quoted in Dumont and Cohen, op. cit., p. 153.

[5] Tony Smith, 'A Comparative Study of French and British Decolonisation', Comparative Studies in Society and History, 20, 1, 1978, p. 92.
[6] George, op. cit., p. 118.
[7] Economist, 15 December 1984.
[8] The African historiography is referred to in Robin Palmer and Neil Parsons (eds), The Roots of Rural Poverty in Central and Southern Africa, London, Heinemann, 1977, pp. 1-2; John Iliffe, A Modern History of Tanganyika, Cambridge University Press, 1979, chs. 9 and 10; and Donald Denoon and Catherine Snowden (ed.), A Time to Plant and Time to Uproot, Boroko, Institute of PNG Studies, 1980.
[9] On the value of the ethnographic data see John H. Moore, 'The Evolution of Exploitation', Critique of Anthropology, 2, 8, 1977, pp. 36-37.
[10] The notion of kinship 'invaded all the field and glutted the analysis of family households, cloaking the concept of relations of production', Claude Meillassoux, Maidens, Meal and Money: Capitalism and the Domestic Community, Cambridge, Cambridge University Press, 1981, Preface. Meillasoux has also noted the stress upon exchange in 'From Reproduction to Production: a Marxist Approach to Economic Anthropology', in Harold Wolpe (ed.), The Articulation of Modes of Production, London, Boston and Henley, Routledge and Kegan Paul, 1980, p. 191.
[11] Considered further in chapter 2 following.
[12] Jean Copans and David Seddon, 'Marxism and Anthropology: A Preliminary Survey', in Seddon (ed.), Relations of Production: Marxist Approaches to Economic Anthropology, London, Frank Cass,
[13] Catherine Coquery-Vidrovitch, 'The Political Economy of the African Peasantry and Modes of Production', in Peter C.W. Gutkind and Immanuel Wallerstein, The Political Economy of Contemporary Africa, Beverley Hills and London, Sage Publications, 1976, p. 94; and Emmanuel Terray, 'On Exploitation: Elements of an Autocritique', Critique of Anthropology, 13 and 14, 4, 1979, p. 35.
[14] There is an interesting discussion of the meaning of surplus in John Pernetta and Lance Hill, 'Subsidy Cycles in Consumer/Producer Societies: The Face of Change' in Denoon and Snowden (eds), op. cit.
[15] Terray, 'Long-distance Exchange and the Formation of the State: the case of the Abron Kingdom of Gyaman', Economy and Society, 3, 3, 1974; see chapter 2 below.
[16] Terray, 1979, op. cit., p. 35; and Meillassoux, in Wolpe, op. cit., pp. 194-197. The exploitation of women among the Kikuyu is discussed in Penelope Ciancanelli, 'Exchange, Reproduction and Sex Subordination Among the Kikuyu of East Africa', The Review of Political Economics, 12, 2, Summer 1980.
[17] Exploitation and class in pre-capitalist societies are discussed in Pierre Philippe Rey, 'Class Contradiction in Lineage Societies', Critique of Anthropology, 13 and 14, 4, 1974, and in Terray, 1979, op. cit. They are also most usefully discussed by de Ste. Croix where exploitation is understood as 'the appropriation of part of the product of the labour of others', and class as 'the collective social expression of the fact of exploitation, the way in which exploitation is embodied in a social structure': and where both these concepts are examined with specific reference to change within non-capitalist modes. Geoffrey

de Ste. Croix, 'Class in Marx's Conception of History, Ancient and Modern', New Left Review, 146, July-August 1984, p. 100.

[18] Moore, op. cit., p. 44.

[19] Terray, 1979. op. cit., p. 38.

[20] David Beach, 'The Shona Economy: Branches of Production', in Palmer and Parsons, op. cit., pp. 55-56.

[21] Peter Geschiere, 'The Articulation of Different Modes of Production: Old and New Inequalities in Maka Villages', African Perspectives, 2, 1978, p. 45; Robin Cohen, 'From Peasants to Workers in Africa', in Gutkind and Wallerstein, op. cit., pp. 155-156; and Charles Van Onselen, 'Worker Consciousness in Black Miners: Southern Rhodesia, 1900-1920', Journal of African History, 14, 2, p. 249.

[22] Ian Roxborough, Theories of Underdevelopment, London and Basingstoke, Macmillan, 1979, p. 73.

[23] Robert Brenner, 'The Origins of Capitalist Development: A Critique of Neo-Smithian Marxism', New Left Review, 104, 1977.

[24] These issues are explored below.

[25] Mark Turner, 'Social Class in Papua New Guinea', Yagl-Ambu, 7, 3, 1980, p. 44.

[26] Rey, op. cit., p. 48.

[27] G.B. Kay (ed.), The Political Economy of Colonialism in Ghana, Cambridge, Cambridge University Press, 1972, pp. 3-37; and Bjorn Beckman, Organising the Farmers, Uppsala, Scandinavian Institute of African Studies, 1976, chapter 1.

[28] Teodor Shanin, 'Peasantry as a Political Factor', in Shanin (ed.), Peasants and Peasant Societies, Harmondsworth, Penguin, 1971, p. 240.

[29] Ibid., p. 238.

[30] The concept of the peasantry was, however, almost completely absent from PNG studies until the 1970s, and Melanesian anthropology preferred to rely on the static and obfuscating notions of clan, kin, and tribe. See Good, 'The Formation of the Peasantry', in Azeem Amarshi, Good, and Rex Mortimer, Development and Dependency: the Political Economy of Papua New Guinea, Melbourne, Oxford University Press, 1979.

[31] Ennew, Hirst and Tribe as quoted in Peter Worsley, The Three Worlds: Culture and World Development, London, Weidenfeld and Nicolson, 1984, p. 71.

[32] Worsley, 'Village Economies', in Raphael Samuel (ed.), People's History and Socialist Theory, London, Boston and Henley, Routledge and Kegan Paul, 1981, pp. 80-85. Further references to Worsley in this section are to this work.

[33] Jonathan Lieberson, 'The Silent Majority', New York Review of Books, 22 October 1981, pp. 34-37. The books considered by Lieberson in this review article are Scott's, The Moral Economy of the Peasant: Rebellion and Subsistence in South-east Asia, and Popkin's, The Rational Peasant: The Political Economy of Rural Society in Vietnam. Further references to Lieberson in this section are to this article.

[34] Goran Hyden, Beyond Ujamaa in Tanzania: Underdevelopment and an Uncaptured Peasantry, Berkeley and Los Angeles, University of California Press, 1980 and No Shortcuts to Progress: African Development Management in Perspective, London, Heinemann, 1983. The peculiarities and the weaknesses in Hyden's approach are more closely considered in the Conclusion below.

[35] Lenin quoted in Roxborough, op. cit., p. 99.

[36] It follows that the much debated and somewhat sterile question of whether the peasantry constitutes a class or a society is only to be meaningfully considered in terms of time and place, the development of modes of production and the inequalities growing within them. With the rise of rural capitalism the significant questions certainly concern the formation of class divisions within the peasantry, and their interactions with each other and, on a differential basis, with non-peasant classes, markets, and the state.

[37] Andre Gunder Frank, Capitalism and Underdevelopment in Latin America: Historical Studies of Chile and Brazil, New York and London, Monthly Review Press, 1967; Keith Griffin, Underdevelopment in Spanish America, London, Allen and Unwin, 1969.

[38] An example of a balanced and sympathetic critique is David Booth, 'Andre Gunder Frank: An Introduction and Appreciation', in Oxaal, Barnett and Booth (eds), Beyond the Sociology of Development, London and Boston, Routledge and Kegan Paul, 1975.

[39] PNG studies have displayed either an unfamiliarity with the dependency notion or an over-concentration upon its shortcomings, e.g. Ross Garnaut, 'The Neo-Marxist Paradigm in Papua New Guinea', and Donald Denoon, 'Does Dependency Theory Illuminate Papua New Guinea's Development?', both in Yagl-Ambu, 7, 3, September 1980.

[40] Roxborough, op. cit., p. 1.

[41] Ibid., p. 68.

[42] Bridget O'Laughlin, 'Marxist Approaches in Anthropology', Annual Review of Anthropology, 4, 1975, p. 359.

[43] For a somewhat different formulation, see ibid., p. 350.

[44] Perry Anderson, Arguments Within English Marxism, London, Verso Editions, 1980, pp. 78-79.

[45] Terray, op. cit., p. 34.

[46] Rey, op. cit., p. 41. Much debate focuses on the concepts of mode of production and social formation, among which is Wolpe, 'Introduction', in Wolpe, op. cit.; Copans and Seddon, op. cit.; Aidan Foster-Carter, 'The Modes of Production Controversy', New Left Review, 107, January - February 1978; John G. Taylor, From Modernisation to Modes of Production, London and Basingstoke, Macmillan, 1979, pp. 105-138.

[47] For Seddon and Copans, French marxist anthropology has displayed 'a kind of university dogmatism', with 'divergent elaborations developed and different interpretations becom[ing] more and more refined', and with an 'unwarranted degree of theorisation'. Op. cit., pp. 34 and 38.

[48] O'Laughlin, op. cit., p. 368.

[49] Anderson, op. cit., p. 98. Anderson's emphasis. He was here referring specifically to the apposite work of E.P. Thompson.

2 Non-capitalist agriculture in the highlands

In the past and present, agricultural production has characterised economic activity for the mass of Papua New Guineans, and the development of agricultural systems constitutes the major force for change. [1] In the last 500 years, two significant shifts in the social formations and their constituent modes of production in the highlands of Papua New Guinea (PNG) can be recognised. [2] The first was the movement from mainly taro producing economies to ones concentrating on sweet potato, a movement which spanned some 300 years over the whole region, and which proceeded unevenly leaving some communities still devoted to hunting and gathering as well as to taro production. Profound as this movement was, it remained restricted to domestic food and fodder production while lifting these to new levels of intensity. The second and bigger shift was the movement in this century from food to commodity production for export chiefly within certain areas of both the highlands and the islands of PNG. This second transformation also proceeded unevenly, but much more rapidly and totally in the areas of concentration, because it was a movement from an orientation upon the domestic to the world economy, and from pre-capitalist to capitalist institutions and social relations. Yet each production system is to be properly understood only in relation to that which preceded it, for each to a differing extent grew out of and built upon the preceding systems. [3]

The highlands social formation towards the height of its independent pre-capitalist trajectory in the early decades of this century appears to have represented a dynamic composite of perhaps three systems of production. It expressed the articulation of an early and by then diminishing hunting and gathering mode of production with a later taro producing system, and with a more recently developing non-capitalist mode appearing in the central regions of intense sweet potato production expressing certain key relations and structures. A patriarchal village economy was developing distinct inequalities and stratification including not only a broad differentiation between female cultivators and male warriors but also the rise of so-called big men. These leaders were basing their power on the exploitation of agricultural production in polygynous households, on their control over women, young men, and bachelors, and over trade and warfare.

The latter activities served towards the reproduction or expansion of the social groups generally, whether household, kin group, or village; social units tended to develop similar production activities and relations in spite of some differences in organisation, size, and ideology. War and trade also served to further the power of leadership particularly. The crucial or determining elements within the developing highlands formation were the exploitation of women's agricultural labour via the ideology and relationships of male dominance most succinctly expressed in the big man. Far from being changeless as has so often been asserted, [4] non-capitalist Papua New Guinean agricultural development led to and was itself promoted by extensive social change in key regions in recent centuries.

If PNG's contemporary economy and society, located within the dynamism of its past, is to be correctly interpreted then the study of its socio-economic history necessarily involves the consideration of the anthropological literature. Anthropology has in a very real sense been social science in the study of the non-Western world, [5] and the discipline has certainly commanded a pre-eminent position in Papua New Guinean studies, as will be considered further below. This major role and contribution has, however, been accompanied by weaknesses of view and method. Anthropology has neglected the development of broad, predictive theory in its rush to record the detail of 'vanishing cultures', and in PNG in general it has been further unbalanced by an especial fascination with the bizarre, and a concentration upon exchange rather than production relations. [6] Melanesian anthropological literature of necessity constitutes the body of source material and data on PNG's past, but it demands a selective and critical reading, for it recognises the saliency of agricultural production and class formation within changing domestic and international environments only hazily and partially.

Early agricultural dynamics

The movement of 'early horticultural people' is supposed to have introduced taro based production into the highlands. It seems possible to be more specific about the direction of this movement than its timing. Use of the new staple possibly came into the Eastern Highlands through the Markham Valley, in the hands of people who were 'either representatives of an agricultural people from the Markham Valley or a fringe Highlands people who obtained taro and knowledge of cultivation from these neighbours'. These 'proto-agriculturalists' moved into the areas which bordered directly upon the upper Markham Valley, Gadsup, Tairora, Kamano, and Agarabi in the vicinity of modern day Kainantu. The diffusion of taro systems within the Eastern Highlands was supposedly down the Lamari Valley to the areas now inhabited by such groups as the Gahuku and Bena near present-day Goroka. [7] Beyond this, some evidence suggests that the 'main migration of agricultural peoples' associated with local developments and innovations occurred progressively from east to west, [8] chiefly down the broad valleys of the central cordillera.

Some prehistorians and archaeologists have placed the origins of agriculture in PNG at a very early period. Bulmer declared in fact that 'intensive agriculture was well established' in PNG about 5,000 years ago,and 'cultivation was widespread' in the highlands before this. [9] The manufacture of stone tools such as axes and adzes would have been a logical accompaniment to the development of extensive agriculture and finished tools were produced at a small number of quarry sites in the highlands. Kafetu was one such location in the upper Asaro Valley of the Eastern Highlands, and tools were traded widely in the highlands 'during the past 5,000 years'. [10] Given the restricted availability of suitable stone, they were also traded well beyond this. The scarce nature of such important manufactures would seem to have presupposed producvtion for trade. [11] The production of objects like stone mortars, pestles and clubheads, was presumably somewhat similar. Trade between the highlands and the coast began, on Susan Bulmer's view, 'at least 9,000 years ago', and long distance trade in a variety of commodities additional to the above, salt, shell, pottery, plumes, etc., is old. [12]

The chronology of production is, however, of less importance than the dynamics of the developing systems. The displacement of hunting and gathering by the new taro practices would have been slow and uneven. Food collecting and hunting would have remained a significant aspect of taro agriculture in certain areas long after it ceased to be so important to others. The Eastern Highlands was generally in the vanguard of change, and therein the vicinities of modern Kainantu and Goroka were first to absorb the taro practice in its broadly westerly movement. But the wide valley corridor down which communications most naturally moved, also served to restrain the development of taro producing systems, and eventually to stimulate within them an important dynamic for change. Waddel has noted the restricted distribution and demanding ecological requirements of taro and yam. The populations using these crops were 'largely confined' to the valley bottoms at altitudes below about 1,700 metres, and were further 'obliged' to establish elaborate drainage works in specific locations such as that in the Wahgi Valley in order to cultivate their staples. [13] The valleys which brought agriculture to particular people, helped ultimately to limit socioeconomic development based on existing agricultural practices. The drainage of the Wahgi Valley was advanced, but such agricultural systems may also be interpreted, said Waddell, as 'having been developed in response to the dictates of the resources exploited, which in turn served to restrict the populations to these areas'. Certain societies in particular locations therefore sought eventual 'release' from 'the constraints imposed by the resources' of the then existing production systems. [14]

The developing overall situation in the highlands was one of 'increasing economic complexity and variation'. [15] Precise identification of developmental trends is or was limited, however, by deficiencies in archaelogical analysis. White emphasised, in 1972, that 'we do not yet know what the artifactual markers of horticulture are', and he referred to the small number of open sites then reported, and to the 'exceptionally small' number of artifacts on which parts of the Bulmers' constructs of the 1960s were based. [16] These problems in archaeological interpretation bring into question the attribution of intensive agriculture to production in the distant past, and suggest that broad theoretical consideration focussed on the moment of systematic change may more fruitfully interpret agricultural development. It was the most developed taro or yam based system in highlands valleys which acquired new agricultural practices of superior productivity and adaptability.

Analogy with societies in the present day suggests that the social relations of the developed taro production systems were characterised by a division of labour based on age and sex, and that their political institutions were controlled by older men. The existence of say specialised axe and pottery making and of long distance trade in certain products, suggests the availability in communities engaged in such activities of a surplus in excess of 'subsistence' needs. It implies a capacity for social inequalities which would tend to deepen given the recurrence and growth of the surplus. [17]

Systems of production slowly developed in parts of the highlands with both need and capacity to absorb sweet potato quickly. [18] Yen has referred to the development of a 'back-ground prepared already agriculturally - in terms of methods and organization - for the expansion that was to ensue'. [19] Developing social relations and organisation were also part of the receptive economic background.

The development of agriculture did not imply the necessary backwardness of hunting and gathering per se. The latter mode was capable of expansion and devlopment in a favourable environment unchallenged by alternative systems of production. The Purari of the Papuan Gulf, for example, experienced notable systemic expansion until the intrusion of colonialism. With an economy based on fishing and gathering in a physical environment of rivers and sago swamps, abundant wild sago palm offered a plentiful, easily obtainable staple by collecting methods. Because sago was also non-seasonal, and the rivers offered seafood to fishers using a technology of dugout canoe, nets, spears and arrows, a 'sedentary and relatively concentrated population' grew up over an unknown number of generations. Both rivers and sago swamps supported the Purari. Without the rivers they 'would have been tied to a pattern of small social groups scattered through the many sago places', but with them and the easy productiveness of sago 'an alternative of large, concentrated communities was added'. [20] Yet this was ultimately the extent of their achievement and their mode of production had little or no capacity to relate successfully to the penetration of the colonial capitalist mode which presented no opportunities to hunting and gathering. Advanced highlands taro producing systems perhaps approached the limits of their development some hundreds of years earlier, but the arrival of the sweet potato offered them a transition to notably higher productivities.

The shift to sweet potato systems

The earliest date for the arrival of the sweet potato on the New Guinea coast is about 1550 according to Yen. The geographic origin of the Pacific plant was the Americas, and O'Brien identified the carrier of the plant. From Polynesia it was transferred to Melanesia by Polynesians, and from these it was carried to New Guinea by Melanesians. It entered the country from the Bismark Archipelago, she suggested, and spread westwards into the highlands through the Markham Valley. [21] This valley and its contiguous areas on the borders of the Eastern Highlands was possibly the busiest of the established trade routes along which new plants and crops were introduced into the highlands. [22] The Arona Gap and Wanton River were almost certainly the main contact routes over a lengthy period for trading between eastern Gadsup and the lower Markham Valley. [23]

In the Eastern Highlands, a 'proto-agricultural movement originating in the Kamano region' in a 'probably recent' period, is credited with the primary role in the diffusion of sweet potato by Sorenson and Kenmore. They noted that Watson found that Agarabi and Tairora people cited Kamano as the area from which the new plant came into their hands. [24] Watson suggested that sweet potato arrived among the Agarabi and Tairora between 200 and 300 years ago, but Sorenson believed that it arrived among the Fore, living close by to the west and southwest, only about a century ago. [25] The diffusion appears to have proceeded at an uneven rate dependent perhaps on the existing levels of agricultural development and on topography in particular places. The Fore were, in Sorenson's view, 'pioneer peoples on the edge of the major Highlands populations', and they pushed gradually into virgin bush. If the Fore's acquisition of the new plant was late in relation to the Kamano, Agarabi and Tairora, it was early by comparison with their eastern neighbours, where sweet potato replaced taro 'only within living memory'. [26]

An uneven rate of diffusion was also found within the highlands region generally. Bowers has suggested that sweet potato arrived among Kaugel Valley people in the Western Highlands about 200 years ago, and Golson has referred to the possible appearance of the planet in the vicinity of Kuk in the Wahgi Valley 250 years before the present. [27] The Southern Highlands, however, showed a lingering attachment to taro as late as the 1960s. [28] If less agriculturally advanced systems in less accessible areas were slow to adopt the new plant, the overall trend towards absorption of the sweet potato in the highlands was very clear. By the time of the colonial explorations of the interior, sweet potato was 'dominant', said Yen, in most highlands areas, and Chinnery reported in a survey of peoples from 'Mt. Chapman to Mt. Hagen' in 1932 that taro and yams were 'not often seen'. [29] What had taken unknown millenia in the development of taro and yam had taken little more than 300 years with sweet potato.

Why the new plant achieved so rapid and wide a dominance in the agricultural systems of the highlands is partly indicated by reference to its productiveness and other special features. Impressive yields of sweet potato have been reported in the highlands in recent decades. Massal and Barrau estimated yields of between 6 and 13 tonnes per hectare and more than 22 tonnes under experimental field conditions; Meggitt recorded nearly 9 tonnes per hectare for the Mae Enga; Brown and Brookfield claimed higher harvests for the Simbu; and Salisbury noted the superiority of sweet potato relative to alternative crops within the Siane economy, recording yields of 9.7 tonnes per hectare for sweet potato but only 0.9 tonnes for taro or yams and 0.7 tonnes for maize. [30] Sorenson also found sweet potato yields among the Fore which were 'much greater than for all the other crops', and he felt that 22 tonnes per hectare in a good garden was common. [31] Sweet potato provided one of the most adequate foods for human consumption from plant sources in terms of bulk, and it possessed undoubted supremacy as fodder especially in comparison with taro or other predecessors in the highlands. [32] Merely in terms of ease of preparation for eating, it was better than alternative Oceanic root or other starch plants. Yet the real advantage of sweet potato lay not so much in the size of its harvests and its advantages as food and fodder, but in its adaptability; as Yen said, 'its ability to yield under a wide range of natural and agronomic conditions'. [33]

Sweet potato's capacity for adaptation represented a crucial linkage with the limitations of the developed taro and yam systems of production on the highlands valley bottoms, and this connection suggests how the shift to the new plant came about. The adoption of sweet potato by proto-Kamano people occurred 'probably under economic and ecological pressures'. [34] At the same time the new plant possessed strong 'pioneering propensities', [35] which fitted well with the altitudinal constraints of the then advanced systems by facilitating population movement to higher levels, from about 1,700 metres to about 2,700 metres according to Waddell. [36] What the articulation of taro and sweet potato systems involved was not a mere change to a better plant and steadily increasing intensification, but a 'conversion', as Watson said, from one whole socioeconomic system to another with definite 'opportunity costs' as well as grave social consequences. [37]

Expansion and intensification

A population movement into higher forest regions has been widely recognised as a unique consequence of the reception of the more adaptable plant. [38] Large numbers of migrants moved from the crowded kunai heartland of the central highlands, and later with the Fore specifically sweet potato permitted the steady movement of gardens and settlements into vast virgin regions at higher altitudes. [39] The impact of the new plant was increased by what Watson and Waddell agree was and is the rapidity with which cropping innovations are integrated into highlands agricultural systems. [40] It was further heightened both by the enlarged yields of sweet potato and by people's increasing reliance upon the new crop in meeting their dietary needs, a dependence reaching levels in recent decades in various highlands communities of between 40 and more than 90 percent of food consumed. [41] The consequence of the arrival of sweet potato in particular societies was therefore not only the geographic expansion of population but plausibly the fairly rapid increase in population numbers too.

Change, however, did not derive from cropping innovations alone. Sweet potato's usage as fodder, combined with the deforestation following the movement of agriculturalists to wooded, higher altitudes, provided a capacity for the growth of domestic pig production. Sweet potato brought about the decline of wild pig, but made pig raising 'possible as well as necessary' in Sorenson's view. [42] Large scale pig production was 'an essential concomitant' to agricultural intensification and population expansion, said Waddell, and it was for Watson 'a main force' in the conversion to the new system, justifying and helping to repay the opportunity costs involved in the process, the abandonment of the impressive earlier drainage works, migration, and so on. [43]

Such expansion and intensification expressed 'the sweep and depth of the conversion', and Watson in consequence referred to the 'ipomoean revolution' in production, demography, and other areas. [44] While the Bulmers in 1964 also pointed to the 'economic and social revolutions which must have followed [the introduction of the sweet potato]', a gradualist perspective was subsequently propounded within Melanesian anthropology and prehistory; the social and demographic effects of the adoption, White said in 1972, 'were probably not very marked in most areas'. [45]

Nelson has, however, challenged the evolutionist view. 'The clue to the pre-contact population increase is ... not to be found in present growth rates', as Brookfield and White implied, 'but in an estimation of what might have been the growth rates <u>before</u> [the introduction and spread of] malaria'. His research in the Nebilyer Valley gave support to Watson's view, for it indicated that a 'population explosion' had in fact occurred in that area during the period of 'about 50 years with sweet potatoes and without malaria' (his emphasis). [46] Waddell concluded that the sweeping change which Watson proposed could well have occurred at altitudes above about 1850 metres. [47]

Changing patterns of warfare and exchange

Increasing population numbers and movements meant increasing interaction among social groups. This was mainly effected via warfare and by trade and exchange wherein agricultural production was the basic element.

Warfare became more frequent and intense in the highlands following the adoption of sweet potato, Watson suggested, because of population growth, the greater propinquity of communities, the pressure upon land, and the greater time and human resources available for war, and Sorenson also saw war as a major consequence of the shift among the Fore. [48] Warfare was directed towards the gaining and destruction of wealth, and was a manifestation of male dominance and a vehicle for the development and consolidation of powerful leadership.

The work of the early ethnographers allows some reconstruction of the situation towards which highlands agricultural systems were tending. In the southwestern vicinity of Kainantu, a state of 'constant insecurity' seems to have existed. [49] With enemies in close proximity, security could be achieved 'only through force'. [50] Attack might be prevented through no passive deterence but through actual, frequent resort to war. The situation was apparently little different in Bena where war was intense and unrestricted. The aims of warfare were 'the complete and total destruction of the enemy ... every man, woman,and child, whether old, infirm, or pregnant ... cases are known in which entire groups were destroyed'. [51] Among the Gahuku-Gama too, warfare was aimed, said Read, at 'the complete destruction of an enemy and his means of livelihood'. [52] There appears to have been in the highlands 'one of the most continuous and violent patterns of warfare on record'. [53]

War had or acquired the specific aims of gaining and destroying wealth. Immovable property was destroyed, villages fired, gardens laid waste, while pigs and produce were stolen. [54] For the people adjacent to Kainantu warfare offered 'a potential source of additional wealth, as well as of corpses for feasting', and the Kuma of the Wahgi Valley fought for pigs and women. [55] Fighting over the key productive resources, land and labour, may have been the particular tendency in areas of most intense sweet potato production. Constraints which affected the old taro and yam systems were again encountered at altitudes where frost became a hazard to sweet potato, once population densities increased. [56] These new limitations were probably much sharper than before, due largely to the increased valuation of land and labour. Warfare in Simbu and Enga, both areas of growing population densities, was often directed towards the conquest and occupation of enemy land. [57] With the Raiapu Enga specifically, encroachments over territorial boundaries were a constant threat, and land disputes were resolved through war. [58]

This was the situation in the most intensive highlands system at the time of white penetration, but it appears to have been derivative of an acceleration and intensification of warfare since the incorporation of sweet potato and its rapid dominance within highlands agriculture.

Warfare was also the expression of the ideal male personality, the leadership type related to Melanesian political-cultural patterns generalised over time and place. This embraced what Berndt termed 'an excess of aggressive and violent action', which was expressed in the related activities of sexual assault, cannibalism and warfare. Treachery was a long accepted part of fighting, and warfare was itself sustained partly through the relatively consistent concentration on male dominance, and on the desirability of using physical violence widely. [59] Fighting ability was a special attribute of the powerful leader, and it was here that ruthless violence was most clearly displayed. While ethnography has tended to minimise the frequency and especially the significance of this leadership form, [60] cases are in fact reported from the Fore, Tairora, Simbu, Siane, and elsewhere.

Matoto of Tairora (died c. 1930) was characterised by audacity and boundless aggression, and Watson noted significantly that there were other leaders in pre-colonial and contemporary Tairora who were not unlike Matoto. [61] Glasse and Lindenbaum found that a big man among the Fore simply 'mould[ed] rules to suit his own needs', and Salisbury stressed, with reference to Simbu, Siane and other areas, that a despot's arbitrariness and terror were 'not merely passively accepted but [were] connived in by supporters, provided that it meant material profits from thefts or other aggression against both group members and outsiders'; somewhat similar to the situation in Tairora, he saw this as a 'recurrent feature' of highlands political structure. [62]

Male aggression and ruthless leadership clearly had quite specific material aims. As Berndt found in Kamano and North Fore in the early 1950s, 'economics and warfare were two sides of the same coin' in the power and action of the warrior leader. [63] If male dominance gave added permanency to war, the nexus with material gain must have grown over time with the increasing valorisation of resources.

Exchange and warfare (leading sometimes to incorporation) were the main means of political accommodation among alien groups during rapid change. [64] Sweet potato production was the basis of these activities, but with the increasingly widespread availability of the crop in the highlands and in the absence of suitable storage techniques, it acquired significant exchange value only when fed to pigs and converted to pigmeat. [65] Domestic pig production had at the same time, according to Watson, a two-fold role in the conversion to sweet potato, encouraging the use of the plant as fodder and stimulating production in general; and through the expansion of pig production 'intensif[ying] the competition among producers for every item or relationship for which pigs [were] essential coin. Among them [were] brides and affines, partners and allies'. [66]

Within Melanesia generally pigs served as a repository for cropping surpluses which were difficult to store. Ceremonial slaughters and feastings served to control pig populations and the demand on plant resources. [67] In the highlands the domestic pig was uniquely suited for exchange and the development of cooperative, competitive relations between different agricultural peoples. Pigs represented considerable value per unit, the amount of labour required to raise them was well understood, and they were an important source of protein. [68] Growing production and trade in stone axes and other tools and materials presumably accompanied agricultural expansion. The Fore, for example, towards the early 1930s, acquired axe blades from groups to the north and west and black palm bows from the south, producing and trading salt, animal furs, feathers, betelnut, cane, and other forest materials themselves. Local Fore groups also exchanged surplus food crops internally, as these matured in different locations at different times of the year. [69] Highlands societies therefore were moving to a stage where, by the time of colonial penetration, sweet potato production provided the economic foundation and warfare and exchange were agencies for competitive intergroup relations.

The expansion of sociopolitical power

Against the anthropological stereotype of the small, fixed descent-based unit, reality was typified by flexibility, change and the incorporation of outsiders under the influence of agricultural development, warfare and competition. Langness pointed to the actuality of the highlands where 'individuals ha[d] a great deal of choice in group membership', and where 'large numbers of non-agnates' were found in many groups, and de Lepervanche emphasised that 'common residence and working together rather than descent-group membership ascribed at birth [were] the bases for group solidarity', in an environment of change and warfare. [70]

The development of production was the key factor promoting change in the social unit in Watson's view. The growth of pig herds and the higher ratio of fodder to forage in the pig's diet constituted pressures to decentralise settlements, while at the same time large scale pig production and distribution provided a capacity for a 'substantial growth of polities'; the massive pig killing ceremonies of, for example, Mt. Hagen and the Wahgi Valley, required large sociopolitical organisations. [71] The dimensions and size of the social unit were affected broadly, said Waddell, by change in agricultural systems, ecology, and defence, and de Lepervanche saw increased agricultural activity and competition leading to efforts 'to widen the executive political unit' and acquire additional security. [72]

Reports on a large number of highlands societies testify to the reality of social expansion under the impact of agricultural augmentation, competition and war. In Gadsup it was the local not the kin group that was of importance to individuals in their daily affairs; in Tairora and among the Tsembaga of the Simbai Valley there was a 'pulsating' flow and residence pattern of outsiders moving into the group and insiders moving out; and among the Kuma 'segmentation, fission, and fusion [were] constantly in process'. [73] Significant numbers of non-agnates came to live with the Gururumba, and the Fore too welcomed non-agnates, offering them inducements to stay and so strengthening the local group. As Glasse said of the South Fore, 'Once a man demonstrate[d] his loyalty to a group by appropriate acts and participation in corporate affairs he belong[ed] to the group in a full sense. The other members care[d] little whether he [was] an agnate, a cognate, or even an unrelated man'. [74] The Simbu likewise saw group membership open to achievement, and in a competitive atmosphere 'developed large groups', as did the Kaimbi of Nebilyer Valley, where in fact the size of social groups 'increased markedly'. [75]

These societies, however, maintained the importance of descent ideology while drawing non-agnates into their midst. With the Kuma the assimilation of outsiders was as it were 'glossed over and hidden', and among the Manga of the Jimi River, Western Highlands, a so called genealogical amnesia was induced to effect an orderly converion. [76] But it was much less the distortion than the adaptation and utilisation of descent dogmas to the exigencies of change and development that was at work. The Tsembaga viewed agnation and territoriality as ideally associated together, and this was a useful notion during a process of continuous residential change and land amalgamation. [77] What was involved was made more explicit by Waddell with reference to the Raiapu Enga, where agnation emerged 'as a mechanism by means of which intensive agriculture [was] possible within the framework of an inherently unstable political system'. Defining descent in terms of territory served to bind loyalties to land, and provided a continuity to agricultural development that was otherwise adversely affected by instability and war. [78]

The incorporation of non-agnates did not occur simply spontaneously but was encouraged by leadership which itself gained strongly from the process. The basis of leadership in the highlands was the control of agricultural production, [79] with intergroup trade and exchange and the plunder from war serving as integral parts of the total system. Among the Raiapu Enga described by Waddell, big men achieved and strengthened their position

> through careful agricultural planning, in which culturally determined limitations on both the availability of and access to open fields lea[d] to major inequalities in levels of production of the staple food within the local group. These, in turn, result[ed] in interhousehold variations in the number of pigs supported, and therefore in the opportunities of adult males for marriage. Consequently, the chances of an individual engaging in ... economically and politically reumerative activities ... [were] also directly affected. [80]

This was the political context in which group segmentation, fission, and fusion occurred, and while the group in general gained from the incorporation of outsiders mainly in precarious, short-term military security, the big man in particular increased the size of his following and thereby the strength of the labour force at his disposal and the surplus production that he controlled.

The sources of power among the Raiapu Enga were not dissimilar to those in other comparable highlands communities. A big man in South Fore was skilled at adding new members to the group, offered refuge to newcomers, and bound them to himself as 'brothers' and 'sisters', and the following of the Gururumba big man LuBiso included immigrant families from Simbu and Gende, attracted by his warrior reknown. [81] The close link between a big man's fighting capability, immigration to the group, and the big man's power was indicated in the case of Tairora and Matoto. The flow of newcomers into Tairora was largely dependent on the actions of a strong leader like Matoto, and the leader gained thereby 'a clear political advantage'. The recruitment of a sizable number of refugees was, said Watson, a 'major political feat' of Matoto, and because of this and his related exchange activities, up to 2,500 people directly, and even more indirectly, 'may sometimes have fought for or thought of themselves as a part of Matoto's enterprises'. [82]

Clearly as a consequence of the adoption of sweet potato in the highlands, there was a consistent, fundamental movement towards the expansion of the social unit in areas of intense production. Movement towards the expansion of the political unit was recurrent, linked as it was to the need to maintain male solidarity [83] in the face of a recalcitrant female agricultural workforce, but less consistent due to widespread warfare and the arbitrary, individual character of the big man's political power.

Changing social relations of production

Both production costs and patterns of work were affected by increasing sweet potato and pig production. Hunting, an extremely energy-intensive form of protein acquisition, was for millenia an activity of the male warrior. [84] The reduction of the forests and the wild game within them, in the areas of most intense production, meant the gradual decline of hunting activity. [85] This occurred at the same time and as a consequence of the fact that pig husbandry offered the alternative of a regular, reliable supply of protein borne upon the availability of a surplus of sweet potato fodder. Pig keeping was the main additional activity of the female sweet potato grower. The intensification of agriculture therefore appears to have involved a notable diminishing of the burden on men along with a significant increase in the labour of women. It 'thrust upon women the major economic burden and an importance greater than before' and there was a rising demand for female labour which must have led to 'heavier work loads for women or to an increasing competition for wives or to both'. [86]

The change in women's work involved, said Watson [87]

> growing larger areas of both staple food and fodder ... transporting ever heavier backloads of tubers sometimes long distances on steep slopes, and ... keeping track of pigs, feeding them, and caring for them.

Situations developed where women planted and tended sweet potato and other crops (mostly not sugar cane, bananas, and yams), harvested most crops, did the heavy work of breaking and preparing the soil, weeded the garden on a daily basis, and gathered firewood. [88] Such labour represented 'drudgery' among the Kuma, where in addition a women might also have to care for, depending on her marital status and her husband's social position, between 15 and 30 pigs. [89] The burden of transportation was itself extreme. A woman returned from the gardens with firewood and food sufficient for her family's evening and morning meals, and with fodder for her pigs. The family for whom she carried and prepared food might total some 10-15 people including adults and children, and the woman might carry loads of food, fuel, and fodder of some 17 kilos or more. [90] Sustained toil of this kind marked a woman physically, as Read noted in referring to the Gahuku woman with 'hollows in her back, large enough to contain my fists ... the stigmata of her lifetime of carrying'. [91]

A substantial sexual division of labour also grew up in manufacturing and building. While men were generally responsible for the construction and maintenance of bridges, and houses which needed rebuilding every three or four years, women commonly performed the 'never-ending task' of transforming vegetable fibres into cord, which was then fashioned into wearing apparel, lashings, pig-tethering rope, and most importantly string-bags or bilums in which all produce was carried. They also wove pandanus mats, obtained and maintained the bamboo cylinders which were used as cooking utensils and containers, and collected grass and reeds for house construction. [92]

Women became increasingly important in production in the sweet potato systems of the highlands under political regimes characterised by, as Reay said of the Kuma, 'male dominance ... manifest in every institutional context'. [93] Suppression, widespread and thorough going, was used to meditate the contradictions, and women were brutally conditioned to accept social inferiority and the fundamental antagonism between themselves and men. [94] Women were subjected to debauchment and serial intercourse in the societies near Kainantu, rape by all males of a lineage was disciplinary procedure for a supposedly defiant wife in the Fore, and female infanticide took place among the Kuma. [95] The mutilation of wives by husbands with sticks, fire, and arrows commonly occurred, and wives were murdered by big men in the highlands both often and widely. [96]

Violence against women was a key aspect of the whole atmosphere of warfare wherein male dominance, as noted, was a prominent feature. But it was also a reflection perhaps of the incomplete effectiveness of the process of suppression given the growing economic importance of women. They did not accept subordination even while they were powerless to change it. Gahuku women suggested this in the way they addressed their husbands. The woman Ilato, Read noted, [97]

> replied in the slightly injured, querelous voice I heard so often that I began to expect it as a woman's natural form of expression. The tone barely recognized the man's right to command and thus asserted the woman's own will even while complying with his demands.

Women also deserted their husbands, [98] a few turned to the hazardous, quasi-independent life of 'wandering women', [99] and physical assaults on men sometimes occurred in ritualised rebellions, for example, at the culmination of an initiation and at a marriage arrangement among the Gahuku, and following the birth of a deformed baby in the Fore. [100] Yet while married women were enjoined to strict fidelity, suppression seems essentially to have focussed on the productive and reproductive role of women. Circumstances of widespread warfare with limited social stratification among men required control of women's labour if accumulation by men within agricultural production was to be achieved. Male dominance and economic development was based on the control of women as sweet potato-pig producer and reproducer of warriors and brides. Pig feasts among the Siane, Salisbury noted, were called by men when the women had 'misbehaved', and they were instructed to show greater 'diligence in gardening and pig care'. [101] The process of control pertained generally to the developing agricultural systems of the highlands. As Langness observed: [102]

> To survive in the New Guinea Highlands, and especially to survive well - to have many large gardens and many pigs - it [was] necessary to control the labour and, indeed, the actual bodies of women. They [had to] do what [was] required of them - and when it [was] required.

Men's labour increased to some extent with sweet potato and the fencing and cutting of new gardens, the clearing of forests in pioneering areas and phases, and the construction of new houses. But the most difficult task which confronted young men in particular was that of obtaining a wife. To be an adult among the Bena

> [was] to be married and have children. A man, in order to be a man, need[ed] gardens and pigs to fulfil his social obligations and to have these he [had to] have a wife to help him. [103]

Marriage for a young man set in train a long series of payments to affinal kin. Such material and social obligations, if they could be entered into and ultimately met were however, the difficult means to a much greater end. While the cost of a wife was considerable, as Watson saw in Tairora, 'the value of a good one clearly exceeded it'. [104]

However, men's labour decreased with sweet potato, not only as noted in the gradual decline of hunting, but also by comparison with the new crop's main alternatives all of which demanded greater labour of men. The crop's wide productiveness made polygyny among the Kaimbi especially feasible and attractive, and represented here and elsewhere in the central cordillera, a powerful dynamic within the division of labour. [105] But the rising value of the female producer saw, not only increasing competition for wives, but also pressure growing on ambitious men to marry additional women. [106] In Tairora women seemed to have been in 'perennially short supply' in terms of wives of the 'right age, appeal, and working and child-bearing capacity' for the men who wanted them. But a polygynous man unquestionably possessed a scarce and valuable resource, and a conspicuous symbol of his masculine attractiveness and social importance. [107] By the time of colonial penetration polygyny was reportedly widespread in the central highlands. [108]

Polygyny inevitably became both a cause and a consequence of leadership and resulted in increasing inequalities among men. The movement of women tended to fall into the hands of those who controlled the flow of goods produced within the agricultural system; this involved circulation within a particular community and trade and exchange between different social groups. Those who controlled exchange were those who dominated sweet potato, and hence pig, production.

As the competition for wives sharpened and the cost of brides heightened, the economic differences between polygynous and monogamous men, not to say unmarried men, widened. [109] The possession of wives necessarily involved the accumulation of pigs and vice versa; only a very wealthy man had more than two wives, and big men were both wealthy and polygynous. [110] Evidence from a wide number of societies shows that polygyny was a causal factor of leadership. Makis, a known warrior and big man of the Gahuku, had 3 wives; LuBiso, a similar figure among the Gururumba, had 12 wives; Kondom of the Simbu had between 7-11 women attached to him and more than 16 children; Pok of the Kaimbi (around 1870-1950) possessed 13 wives and 38 children; and Kavagl, (sometimes, Kauwagl) another Simbu leader, born about 1890, had wives sufficient for him to kill two in fits of anger. [111]

Large scale polygyny tended to be the particular attribute of powerful, ambitious leadership. Pok, according to Nelson, progressed by managing women's labour in the development of sweet potato. He

> apparently acquired a wife or two through traditional means, and invested their labour in sweet potatoes to the extent that he could manipulate food resources. This brought him retainers and control over circulating wealth, more wives, and an escalating cycle of prestige, authority, wealth, sons, and control over the bridwealth of daughters. [112]

Pok was in fact 'the first of a new breed of entrepreneurs' at work in the period following the introduction of sweet potato and before the imposition of colonial control over the Kaimbi. But his achievement was greater than that of contemporary Kaimbi big men not least because none of these have had more than 7 wives. [113] In Tairora the despot Matoto had 16 or more wives. The economics of Matoto's position were 'centrally focussed' on the number of these women and his command over their production. The wealth which came to powerful leaders, Watson concluded, 'took the form notably of women in marriage, and ... could hardly have taken a more meaningful economic form' under the prevailing circumstances. [114]

Women nevertheless represented certain important political problems to highlands men in general and to leaders specifically. Wives, which most women were actually or potentially, came from outside the community and owed loyalties to enemies or potential enemies. The integration of new settlers obviously raised similar problems of allegiance and trustworthiness for leadership. The only sure method of gaining new warriors was therefore natural increase, but dangers existed for male dominance in women's necessary influence over the early upbringing of male children. Boys in the Eastern Highlands were there- fore forcibly removed from their mothers and the company of other females on a number of occasions. [115] At the earlier separations the boys were physically marked, later they underwent violent experiences to test and strengthen their sense of male superiority and unity, and then at puberty they were removed to the men's house where they resided thereafter. Young men were continually warned of the supposedly dire consequences of association with women, basically that they would become 'rubbish men' rather than 'men with a name'. [116] Male antagonism was perhaps partially overcome (and expressed) in sexual intercourse, [117] but here too was danger, for women threatened the vital numerical strength of the group when they controlled their fertility. Read noted a considerable variation in fertility among Gahuku women, with some having had as many as six children, but with the average only two, and Langness found it was widely known among the Bena that women sometimes prevented conception or induced abortion. [118]

Control over their fertility may have been women's attempt to balance their burdensome mothering and productive roles in circumstances characterised by their near total powerlessness. Men exercised ultimate control in agricultural production and owned the key tools and all weapons. Among the Gururumba women used land dependent on their marital and family positions, but they had no rights of alienation over land or over houses or forests. [119] A Mt. Hagen woman could make small gifts of sweet potato when her husband's needs had been adequately met, but this limited freedom probably differed little from the situation in the Gururumba where men 'exercise[d] ultimate control over land and the usable products on it'. [120] Hagen women disliked men appropriating their pigs for trade and exchange [121] when they cared for the animals, suckling them sometimes at their own breasts. Women owned digging sticks, but the axe was a man's tool and was the instrument which produced the more ubiquitous digging stick. [122] Axes as noted were commonly obtained through trade, an activity controlled by men. Weapons were similarly the preserve of men, and among the Gururumba it was an offence for a woman even to touch a man's bow. [123] Women exercised final control usually over only their digging implements, ornaments, and the <u>bilums</u> important in transportation and clothing. Their real position of economic importance combined with political powerlessness was expressed in the Fore saying: 'Wives do all the hard work, they are the hands of men'. [124]

The exploitation of women's labour by men generally and by big men most particularly was the conspicuous factor in highlands' social relations. With continuing agricultural development the mobilisation of both male and female labour became increasingly important in areas of intensive production. While questions concerning production and labour were usually not the prime concern of the early anthropologists, some impression can still be gained of the changing social relations. Only big men could command labour among the Kuma, and migrants from the densely populated Simbu valleys were employed as garden labourers in Susuroka village near Goroka in the early 1950s on the eve of the introduction of coffee. [125] Vicedom and Tischner's account of Hagen social structure and change was, however, unusually detailed and related to the moment of initial colonial penetration into the Western Highlands, in other words, to the end point of independent agricultural development in the region. They identified four levels of social stratification. [126] At the top were the nuim or big men and major big men. Next were the ketl, a kind of middle social category who were independent of the nuim though they sometimes worked for the big men and helped them in their enterprises. Below these were 'rubbish men' and 'poor men' who had no independent means of support and were therefore dependent on the nuim for whom they worked as servants. The lowest category, which overlapped with that of rubbish men, was the wagen or bachelor men who worked for the wives of the nuim as swineherds and were therefore little more than slaves.

The power of the Hagen big men was manifest in their influence over inter-group ceremonies and domination over trade and exchange, and its basis was their control over the labour of wives, servants, and bachelor-slaves. A big man was known for the ruthless, if necessary, violent pursuit of his own ends, and by the induction of outsiders into the group as dependent workers. Nuim, moreover, were drawn only from certain families. 'Rich men ensure[d] that their sons [were] endowed with wives, land, pigs, and valuables, so that they inherit[ed] their father's status', though not necessarily his particular eminence since all sons were rivals one of the other. [127] Class formation in the Hagen social situation was more than embryonic and was based largely upon the exploitation of labour.

Highlands agricultural development involved widening inequalities between men and women, the polygynous and the monogamous, the married and the unmarried, and the controllers of labour and other resources and the powerless. There was increasing dependency by young men on big men for brides and by new settlers on big men for protection so that their relegation to the category of rubbish men might be avoided; as Vicedom and Tischner said of the Hagen social trajectory, 'Whoever remains a bachelor becomes a slave'. [128] Changing social relations involved not a balanced reciprocity of patronage extended for support received but the mobilisation of labour and the accumulation of wealth by leadership. Rising inequalities in access to agricultural resources did not lead to meaningful measures of social compensation, but to widening social inequalities and the enrichment of big men. As Waddell said with reference to the Raiapu Enga. [129]

> the solution which the powerful offer[ed] the impotent, of incorporating them in their households, [was] a measure that largely benefit[ed] the former. The 'big man' thereby increase[d] the size of the labour force at his disposal and the surplus production that he directly control[led], in return for protecting the 'rubbish man' from likely public ridicule and isolation.

Rising powerful leaders were themselves an active and originating force for agricultural development and social change. The intensification of production, warfare, and exchange together represented over time a firm 'strain towards entrepreneurship'. [130] Pok and Matoto clearly exemplified such thrust and were not alone, either in their own areas or elsewhere under intensification, in so doing. Their actions were directed towards expansion and accumulation and against those whose position was variously opposed to the drive - chiefly women, young men, bachelors. This in turn ultimately embraced, as the Hagen situation of around 1930 suggested, action in favour of the formation of what was in the process of becoming a nascent class certainly of big men but perhaps of hereditary chieftains. Thrusting big men were 'pivotal' [131] in the intensification of non-capitalist agricultural systems in the highlands and at the root of class formation in the region.

Anthropology and highlands development

Socioeconomic change was notable in recent centuries in the highlands, and the gradualist perspective which would have it otherwise has merely reaffirmed the vision of the colourful mosaic, and served to 'perpetuate the engima' of the region, [132] source of endless variety and wonder. The weaknesses of this position are those of Melanesian anthropology which has generally tended to minimise change, its agency, and consequences. Before the 1960s ethnologists in the main had not considered development in the region and mostly preferred to under- estimate the rate of change rather than the opposite. [133] Wide and continuing reliance on the concept of the ethnographic present introduced a strong ahistorical bias and an air of unreality into much of the literature. [134]

The neglect of production has been accompanied by a strong emphasis on exchange as the key socioeconomic activity and the basis of political power. Strathern said in 1979 that 'big-men still essentially rely on influence over exchanges ... as their basic strategy', and when he spoke of 'home production', he stressed the limitations upon its continuing expansion. [135] Sillitoe allowed that a big man's wives increased both the labour and the pigs that he controlled, but he recognised power only in the indirect form of reknown and reputation, dependent specifically 'not upon the production, but upon the exchange of wealth'. [136]

Weak leadership and the small, fixed social unit were the perceived accompaniments to and consequences of systems characterised by negligible change and determinant exchange networks. Equal opportunity and indeed 'consensus democracy' were the basic principles of the Hagen Mbowamb for Brandewie, [137] if not for Vicedom and Tischner earlier. Sillitoe saw leaders as having no authority over others in Melanesia, and confronted by the reality of a Kavagl and a Matoto, dismissed them as 'only bullies without any influence ... whose outlandish stories people remember and embellish upon because they are so atypical'. [138]

Production and leadership were, however, closely and dynamically linked. This was seen in the rise of sweet potato production and may be deduced from agricultural development generally. Agricultural production presupposes a cycle of interconnected activities with results that are constantly deferred. Privileges of seniority arise in these conditions, as newcomers and young men become dependent upon elders for nourishment, instruction, and access to resources. [139] Big men and production would seem to have been somewhat interrelated in the highlands of PNG allowing for the differences between storable grain crops and perishables like sweet potato.

Ideology served well to disguise the exploitation practised by the big men. Beliefs in patrilineal descent supported the assertion of male supremacy and unity, [140] and powerless men were thereby co-opted through their superiority to women and bound by it to their better off leaders. [141] Such ideology operated under very favourable conditions over much of highlands history, for exploitation of agricultural producers categorised by age and sex was more easily presented as part of the natural order of things. [142] But it became less easily hidden in the period of advanced sweet potato production, when accumulation of resources by big men grew more obvious, naked despotism occurred, and a definite stratification appeared in regions of the central cordillera.

If anthropology has enjoyed a general preeminence among the social sciences of the non Western world, anthropology in Melanesia has been characterised by its special relationship with Australian colonialism. Anthropologists were present very early in the colonial period in the highlands,and their initial activities were unencumbered by missionaries, explorers and planters, the competitors of comparable situations. [143] Anthropology and colonialism in PNG were drawn together literally by the advances of the science. Malinowski and Radcliffe-Brown were responsible for the rise of functional anthropology after World War One, [144] and they, according to West, 'made social anthropology relevant to colonial administration'. [145] Malinowski distinguished the colonial official from the trader, miner, and the other colonial categories, and strongly praised the supposedly 'practical', 'impartial', and knowledgeable colonial administrator. Anthropology, he advocated, 'ought to be made to collaborate' with the colonial government in what he explicitly saw as a subordinate and supportive role. The science could be 'useful in an advisory capacity', and anthropologists could 'consult' with the official on the spot, 'provided' they were interested in 'present-day conditions', in 'realities' such as 'the process of racial co-operation and mixture, and of cultural assimilation'. [146]

In 1923 resolutions were passed at the Pan-Pacific Science Congress in Melbourne, under the presidency of A.C. Haddon, Cambridge anthropologist and exponent of scientific field work, calling on the Australian government to appoint an official anthropologist in New Guinea, to contribute to the establishment of a chair of anthropology at Sydney University, and to introduce a cadet field officer system with officials taking courses in anthropology. [147] Radcliffe-Brown, himself a senior colonial administrator in Tonga, 1916-1919, became first professor of anthropology at the University of Sydney in 1926, and introduced courses of study in social anthropology for cadet officers. [148] The actual extension of colonialism in the central highlands was carried out by some of these university trained officials, who included in their number such active and subsequently prominent figures, as A.A. Roberts, I. Downs, G. Greathead, and T. Aitchison. [149] The courses in social anthropology, said Fortes in 1949, 'remained an important part of the work of the Sydney School of Anthropology as a link between academic anthropology and practical administration'. [150]

Scientific researchers recommended effective support for colonialism with much enthusiasm. Before he ended his tenure at Sydney in 1931, Radcliffe-Brown spoke quite explicitly about the relationship between anthropology and colonialism. What he wished to see in New Guinea was 'a body of research workers engaged in the systematic study of the native social organization handing on their knowledge to an administration that would be able to decide on policies to be adopted'. [151] William Groves, in 1935, saw anthropology as concerned with 'the reconciliation of natives to the changes which must inevitably take place as a result of European contact'. He especially proposed the establishment of a Native Affairs Field Staff, composed of a small number of officials with wide anthropological training. They were supposed to 'concentrate upon the business of directing and adjusting the lives of the natives in their villages'. They would, for example, become '*au fait* with movements ... that the people were incorporating into their new lives', and would 'get to know personally influential natives and leaders'. [152]

Colonial bureaucracy also brought anthropology into its functions and outlook. E.W.P. Chinnery was in the Papuan administration when he received correspondence and 'literature on native races' from A.C. Haddon. Thanks to this, Chinnery was able to study for two years as a research student under Haddon and W.H.R. Rivers at Cambridge. [153] He was appointed Government Anthropologist in New Guinea in 1924, and in 1932 he was named as the first Director of District Services and Native Affairs while retaining his post of Government Anthropologist.

Chinnery was inclined to believe that an extremely lengthy period, implicitly an indefinite one, would necessarily lapse before 'New Guinea [could] be known anthropologically' through intensive study of single groups. [154] He thought that the progress of the colonised peoples was 'inextricably linked with that of European enterprise', and he was particularly keen to direct anthropological research towards the spheres of supposed cultural adjustment and assimilation. [155] The issue of the adjustment of the colonised to capitalist enterprise was closely related to the aspects and uses of anthropology which had most interested Chinnery and others in the first place. The insights which the science provided had in fact enabled Chinnery 'to use new methods', which brought him 'into closer and more intimate touch with native thought ... which took [him] further into the confidence of the people themselves', and placed the 'reins of control ... in his hands'. [156]

University training and contacts acquainted some important colonial officials with the anthropological perspective. It helped to produce a similarity of outlook between official and scientist, and 'sharpened the edge of dislike', said West, between colonial government and missions. [157] Another senior figure who combined the roles of official and scientist was F.E. Williams, long serving Government Anthropologist in Papua and author of a number of ethnographic monographs. [158] Williams' work won the admiration of A.P. Elkin, the successor to Radcliffe-Brown at Sydney. [159] All such associations and cohesions did not, however, alter the inherent imbalance in the relationship between colonialism and science, for colonial government rendered anthropology the fundamental service of making the object of anthropological study accessible and safe for the scientific field- worker. [160]

Despite the lateness of both colonialism and anthropology in the highlands, beginning when a body of anti-colonial ideas and experience was rapidly increasing, Melanesian anthropology was largely uncritical of the political situation in which it operated. Anthropology's basic emphases on changelessness, egalitarianism, and uniqueness were intrinsically supportive of a colonial system which stressed the gradualism, altruism, and uniqueness of its own role. The ethnographer's unchanging colourful mosaic was little different from the political subjectivism of Paul Hasluck, Australian Minister for Territories, who in November 1951 declared: 'New Guinea is neither a colony nor a territory; it is in the experimental stage of something which the world has not yet seen and which it may not be possible to create anywhere else in the world ...' [161] Anthropology's association with colonialism represented a significant barrier against a scientific investigation of production and exploitation in the highlands and a reinforcement of the static perspective.

The limitations of the advanced sweet potato systems

The advance of the sweet potato systems was achieved through the exploitation of land, labour and the products of the land, with only a low development of productive forces, simple stone and wooden tools, no metalwork, no wheel, etc. and with surpluses dissipated in warfare and used in direct consumption. Further to this, the sharp constraints present within the new systems ultimately emerged, chiefly in the form of high population densities and land shortages, in precisely the most intensive areas, namely, the higher altitudes of Simbu and Enga. [162] The attendant social pressures led to particularly high incidences of warfare; among the central Simbu, said Meggitt, a group's desire 'to gain and to hold arable land has been the most powerful motive impelling them to make war on each other'. [163] They also led of course to the migration of population, and eventually to what became in the case of Simbu an outflow of labour to the advanced but somewhat less intensive systems in the Eastern Highlands; Simbu as noted were employed as paid labourers by indigenous farmers near Goroka about 1950. [164]

Pre-capitalist highlands agriculture developed to an extent that won deserved admiration from the first white explorers in the early decades of this century, but costs underlaid these gains. Heavy reliance on a single crop was itself insecure, with frosts and droughts capable of causing famine in particular areas such as central Enga where sweet potato was grown near the upper altitudinal limits of the plant. [165] Optimal conditions for highlands agriculture became in fact 'rare', and Brookfield noted the 'widespread occurrence' of almost marginal conditions, and the premium placed thereby on control of the best land. [166] If Simbu and Enga were precariously placed regarding land and population as highlands agriculture reached the end of its independent development, most areas faced limits on the production of their staple, while some, like those near the valley of Goroka in the Eastern Highlands, could sell land and profit from the inflow of potential wage labour. [167]

The colonial forerunners in the highlands, the missionaries and prospectors in particular, entered by way of the old trade and communications route of the Markham Valley to what became Kainantu, Goroka, and Mt. Hagen. [168] The early colonialism recognised the rich valleys as suitable for white settlement and also noted 'the availability of a teeming potential labour force'. [169] But the articulation of colonial capitalism and highlands pre-capitalist agricultural production took place in such a way as to differentiate intra-regionally between whence the labour was supplied and whence the plantations were established. The advanced highlands sweet potato systems had a generalised capacity, both materially and socially, typified by the thrusting, entrepreneurial big man, for a second transformation to a richer, more developed system. But the incoming foreign capitalist agriculture discriminated between highlands areas. [170] The administrative centres, new services, the roads, and after 1950 the foreign and later indigenous coffee plantations, were mainly established first in the Eastern and then the Western Highlands, while Simbu was firmly relegated to the position of labour reservoir for the islands and highlands plantations. Some Eastern Highlands societies gained immediate benefits to weigh against the disruption and the loss of their independence. Colonialism provided the Siane, around 1933-1945, with a new and expanding market for surplus food production, steel axes facilitated men's work, and growth enabled considerable political consolidation to occur. [171] Other societies adjacent to the main communications route of the Eastern Highlands gained similar early advantages, but the full repercussions of the incorporation of production for commodity exportation took some time to appear.

Notes

[1] Eric Waddell, The Mound Builders, Seattle and London, University of Washington Press, 1972, pp. 6-10. Waddell was referring specifically to the highlands, but the view has wider significance in PNG. It has underlaid a number of African studies, e.g., Robin Palmer and Neil Parsons (eds), The Roots of Rural Poverty in Central and Southern Africa, London, Heinemann, 1977, and constitutes a theme in the history of agriculture in PNG, edited by Donald Denoon and Catherine Snowden, A Time To Plant and a Time to Uproot, Boroko, Institute of PNG Studies, 1980.

[2] The highlands may be considered as a distinctive cultural and linguistic province on the basis of a precise definition of the region emphasising, following Brookfield, distinctive agricultural practices ('certain techniques of land use'). R.M. Glasse and Shirley Lindenbaum, 'The Highlands of New Guinea: a Review of the Ethnographic and Related Problems', in R.W. Hornabrook (ed.), Essays on Kuru, Faringdon, Berks, Q.E.W. Glassey Ltd., 1975, p. 6.

[3] This approach remains necessary despite the fact noted by Heider in 1967 that much of the past may have vanished without leaving any trace. Referred to in ibid., p. 8.

[4] Thus, the statement: 'In the stateless societies of Melanesia ... life follows a constant pattern with no appreciable change'. Paul Sillitoe, Give and Take: Exchange in Wola Society, Canberra, A.N.U. Press, 1979, p. 111.

[5] Robert F. Maher, New Men of Papua, Madison, University of Wisconsin Press, 1961, p. 6.

[6] Maher, op. cit., p. 7, and Paula Brown, 'Non-Agnates Among the Patrilineal Chimbu', Journal of the Polynesian Society, 71, 1962, p. 57. One example of the concern with the bizarre (and exchange) is Gilbert H. Herdt, 'Semen Transactions in Sambia Culture', in Herdt (ed.), Ritualized Homosexuality in Melanesia.

[7] E. Richard Sorenson and Peter E. Kenmore, 'Protoagricultural Movement in the Eastern Highlands of New Guinea', Current Anthropology, 15, 1, 1974, pp. 67 and 72.

[8] Susan and Ralph Bulmer, 'The Prehistory of the Australian New Guinea Highlands', American Anthropologist, 66, 1964, p. 51.

[9] Susan Bulmer, 'Settlement and Economy in Prehistoric Papua New Guinea: A Review of the Archaeological Evidence', Societe des Oceanistes, 31, 46, 1973, pp. 67 and 45.

[10] Ibid., p. 13.

[11] Brown suggests this but on the anthropologist's timeless perspective. Paula Brown, Highlands Peoples of New Guinea, Cambridge, Cambridge University Press, 1978, p. 43.

[12] Susan Bulmer, op. cit., p. 24.

[13] Waddell, op. cit., p. 219.

[14] Ibid.

[15] Susan Bulmer, op. cit., p. 43.

[16] J. Peter White, Ol Tumbuna, Canberra, A.N.U., Department of Geography, 1972, pp.144-146. It is also notable that highlands garden culture has been dated to approximately 350 B.C. on the basis of what was apparently only a single wooden digging stick found in a ditch in swamps near Mount Hagen. R.G. Robbins, 'Vegetation and Man in the South-West Pacific and New Guinea', in R.G. Ward (ed.), Man in the Pacific Islands, Oxford, Oxford University Press, 1972, p. 88.

[17] See the interesting consideration of this point in John Pernetta and Lance Hill, 'Subsidy Cycles in Consumer/Producer Societies: The Face of Change', in Denoon and Snowden, op. cit.

[18] 'Needs' in the sense of a requirement for and a drive towards expansion of the economic system, in this case of agricultural production. Usage of this kind is referred to in Barbara Bradby, 'The Destruction of Natural Economy', Economy and Society, 4, 2, 1975, p. 148.

[19] D.E. Yen, The Sweet Potato and Oceania: An Essay in Ethnobotany, Honolulu, Bernice P. Bishop Museum Press, 1974, p. 322.

[20] Maher, op. cit., pp. 15-18.

[21] Yen, op. cit., p. 324, and Patricia J. O'Brien, 'The Sweet Potato: Its Origin and Dispersal', American Anthropologist, 74, 1972, pp. 360-361.

[22] Michael Alpers and D. Carleton Gajdusek, 'Changing Patterns of Kuru: Epidemiological Changes in the Period of Increasing Contact of the Fore People with Western Civilization', American Journal of Tropical Medicine and Hygiene, 14, 5, 1965, p. 855.

[23] E. Young, The People of the Upper Ramu: A Socio-Demographic Survey of Agarabi-Gadsup, Department of Geography, UPNG, Occasional Paper No.8, December 1973, p. 18.

[24] Sorenson and Kenmore, op. cit., p. 72.

[25] Watson, 'The Significance of a Recent Ecological Change in the Central Highlands of New Guinea', Journal of the Polynesian Society, 74, 1965, p. 440; and Sorenson, 'Socio-Ecological Change Among the Fore of New Guinea', Current Anthropology, 13, 3-4, 1972, p. 349.

[26] Sorenson, The Edge of the Forest: Land, Childhood and Change in a New Guinea Protoagricultural Society, Washington, D.C., Smithsonian Institution Press, 1976, pp. 17 and 79.

[27] Bowers' estimation is referred to in Watson, 'Significance of a Recent Ecological Change ...', p. 440; Jack Golson, 'No Room at the Top: Agricultural Intensification in the New Guinea Highlands', in J. Allen, Golson, and R. Jones (eds), Sunda and Sahul: Prehistoric Studies in Southeast Asia, Melanesia and Australia, London, New York, San Francisco, Academic Press, 1977, p. 626; and H.E. Nelson, 'Disease, Demography, and the Evolution of Social Structure in Highland New Guinea', Journal of the Polynesian Society, 80, 1971, p. 208.

[28] Watson, 1965, op. cit., p.440.

[29] Yen, op. cit., p. 324; and E.W.P. Chinnery, 'New Guinea', Man, 24, August 1934, p. 117.

[30] Yen, op. cit., p. 50.

[31] Sorenson, 1976, op. cit.

[32] Yen, op. cit., p. 49; and Watson, 'Pigs, Fodder, and the Jones Effect in Postipomoean New Guinea', Ethnology, 16, 1, 1979, p. 60.

[33] Yen, op. cit., pp. 57 and 51.

[34] Sorenson and Kenmore, op. cit., p. 72.

[35] Yen, op. cit., p. 116.

[36] Waddell, op. cit., p. 219.

[37] Watson, 1979, op. cit.

[38] Yen, op. cit., p. 116.

[39] Watson, 1979, op. cit., p. 59; and Sorenson, 1976, op.cit., p. 79.

[40] Waddell, op. cit., p. 10.

[41] Yen, op. cit., p. 45; and H.C. Brookfield, 'The Ecology of High-

land Settlement: Some Suggestions', American Anthropologist, 66, 1964, pp. 22-24.

[42] Sorenson, 1976, op. cit.

[43] Waddell, op. cit., p. 210; and Watson, 1979, op. cit., pp. 60 and 64.

[44] Watson, 1979, op. cit., p. 68; and 1965, op. cit., pp. 443-445.

[45] Susan and Ralph Bulmer, op. cit., p. 52; Brookfield and White 'Revolution or Evolution in the Prehistory of the New Guinea Highlands', Ethnology, 7, 1968; and White, Ol Tumbuna, p. 148.

[46] Nelson, op. cit., pp. 206-209.

[47] Waddell, op. cit., p. 208.

[48] Watson, 1965, op. cit., p.447; Sorenson, 1976, p. 93.

[49] Ronald M. Berndt, 'A Cargo Movement in the Eastern Central Highlands of New Guinea', Oceania, 23, 1, 2, 3, 1952-53, p. 44. Berndt was located in Usurufa with Kamano, Jate (or Yate) and North Fore in the vicinity.

[50] Berndt, Excess and Restraint, Chicago and London, University of Chicago Press, 1962, p. 177.

[51] L.L. Langness, 'Traditional Political Organization', in Ian Hogbin (ed.), Anthropology in Papua New Guinea, Clayton, Melbourne University Press, 1973, pp. 162-163.

[52] Quoted in ibid., p. 162.

[53] Langness, 'Some Problems in the Conceptualization of Highlands Social Structures', American Anthropologist, 66, 4, 2, 1964, p. 174.

[54] Berndt, 1952-53, op. cit., pp. 44-45, and 1962, op. cit., p. 259.

[55] Berndt, 1952-53, op. cit., p. 47; and Marie Reay, The Kuma: Freedom and Conformity in the New Guinea Highlands, London and New York, Cambridge University Press, 1959, p. 22.

[56] Waddell, op. cit., p. 219.

[57] Brown, op. cit., p. 206.

[58] Waddell, op. cit., p. 186.

[59] Berndt, 1962, op. cit.

[60] Watson, 'Tairora: The Politics of Despotism in a Small Society', in Berndt and Peter Lawrence (eds), Politics in New Guinea, Nedlands, W.A., University of Western Australia Press, 1971, pp. 231 and 274; and Paula G. Rubel and Abraham Rosman, Your Own Pigs You May Not Eat, Canberra, A.N.U. press, 1978, p. 293.

[61] Watson, ibid., pp. 258 and 273.

[62] Robert Glasse and Shirley Lindenbaum, 'South Fore Politics', in Berndt and Lawrence, op. cit., p. 376; and Richard F. Salisbury, 'Despotism and Australian Administration in the New Guinea Highlands', American Anthropologist, 66, 4, 2, 1964, pp. 228 and 236.

[63] Berndt, 'Political Structure in the Eastern Central Highlands of New Guinea', in Berndt and Lawrence, op. cit., p. 390.

[64] Sorenson, 1972, op. cit., pp. 350 and 360.

[65] Waddell, op. cit., p. 189.

[66] Watson, 1979, op. cit.

[67] Yen, op. cit., pp. 52-54 and 116-117. He refers to the Kaugel Valley as the single example of sweet potato storage in the New Guinea Highlands, and offers comparisons with the systematic storage systems of the Philippines and the Maoris.

[68] Sorenson, 1976, op. cit., pp. 76-77.

[69] Shirley Lindenbaum, Kuru Sorcery: Disease and Danger in the New Guinea Highlands, Palo Alto, Mayfield, 1979, p. 77.

[70] Langness, 1973, op. cit., pp. 149 and 153; and Marie de Lepervanche, 'Descent, Residence and Leadership in the New Guinea

Highlands', Oceania, 38, 2, 3, 1967-68, p. 143.

[71] Watson, 1979, op. cit., pp. 61-65.

[72] Waddell, op. cit., p. 9, and de Lepervanche, op. cit., p. 157 where she was quoting Ronald Berndt.

[73] Brian M. du Toit, 'Filiation and Affiliation Among the Gadsup', Oceania, 35, 2, 1964, p. 88; Watson, 'Society as Organized Flow: The Tairora Case', Southwestern Journal of Anthropology, 26, 2, 1970, p. 110; Roy A. Rappaport, Pigs for the Ancestors: Ritual in the Ecology of a New Guinea People, New Haven and London, Yale University Press, 1967, p. 21; and Reay, op. cit., p. 32.

[74] Phillip L. Newman, Knowing the Gururumba, New York, Holt, Rinehart and Winston, 1975, pp. 6 and 59; and Robert Glasse, 'Cannibalism in the Kuru Region of New Guinea', New York Academy of Sciences Transactions, 2, 29, 1967, p. 750, and quoted in Sorenson, 1972, op. cit., p. 382.

[75] Paula Brown, 'Non-Agnates Among the Patrilineal Chimbu', Journal of the Polynesian Society, 71, 1962, pp. 59 and 62; and Nelson, op. cit., p. 218.

[76] Reay, op. cit., p. 36; and Edwin A. Cook, 'On the Conversion of Non-Agnates into Agnates Among the Manga, Jimi River, Western Highlands District, New Guinea', Southwestern Journal of Anthropology, 16, 1970, p. 195.

[77] Rapapport, op. cit., p. 24.

[78] Waddell, op. cit., pp. 215-216.

[79] Ibid., p. 9.

[80] Ibid., pp. 216-217.

[81] Glasse and Lindenbaum, op. cit., p. 320; and Newman, op. cit., p. 59.

[82] Watson, 1970, op. cit., p. 115; 1971, op. cit., pp. 240 and 253. Not dissimilarly there is mention of an unnamed highlands 'fight leader' who 'could call out a thousand bowmen', in W.G. Sippo, 'Pushing Back the Frontier', Australian Territories, 2, 4, 1962, p. 6.

[83] The men's house was 'a nerve centre and a clearing house for the affairs of a settlement', Watson, 1971, ibid., p. 251; it provided a formal institutional focus for political activity and the maintenance of male supremacy and it was there the key decisions were taken on ceremonies, war, migration, and so on. But the big man occupied the pre-eminent position within it. While the men's house disappeared with colonial penetration, supplanted by the 'informal court', big men survived and indeed in many instances prospered in modern political organisational activity and in commercial agriculture. The men's house is considered further in M. Donaldson, 'Contradiction, Mediation and Hegemony in Pre-capitalist New Guinea: Warfare, Production and Sexual Antagonism in the Eastern Highlands', pp. 435-437 in Melanesia: Beyond and Diversity, Vol. 2, R. May and H. Nelson (eds), Research School of Pacific Studies, Australian National University, Canberra 1982.

[84] W.T. Divale and M. Harris, 'Population, Warfare and the Male Supremacist Complex', American Anthropologist, 78, 3, 1976, p. 524. An experiment among the Siane Komonku indicated that 25.1 man-hours were expended for each kilogram (bag weight) of possum, cuscus, or rat obtained, and more than 5 man-hours were expended for each kilo of flying fox; P.D. Dwyer, 'The Price of Protein: Five Hundred Hours of Hunting in the New Guinea Highlands', Oceania, 44, 1974, pp. 278-293.

[85] By the 1960s, hunting had become 'unimportant' in Bena, and con-

tributed little to people's material needs even in more peripheral South Fore. L.L. Langness, 'Marriage in Bena Bena', and R.M. Glasse, 'Marriage in South Fore', in Glasse and M.J. Meggitt (eds), Pigs, Pearlshells, and Women: Marriage in the New Guinea Highlands, Englewood Cliffs, N.J., Prentice-Hall, 1969, pp. 39 and 17, respectively.

[86] Watson, 1965, op. cit., p. 445; 1979, op. cit., p. 67.

[87] Watson, 1979, ibid., p. 67.

[88] Langness, 'Ritual, Power, and Male Dominance', Ethos, 2, 1974, p. 191.

[89] The typical herd of a big man's wife. Reay, op.cit., Preface and p. 95.

[90] Women provided food averaging 16.8 kilos in Moke Village in North Fore in 1957, at a time when most of the young men were in jail for fighting, L.H. Reid and D.C. Gadjusek, 'Nutrition in the Kuru Region II: A Nutritional Evaluation of Traditional Fore Diet in Moke Village in 1957', Acta Tropica, 26, 4, 1969, pp. 334-337; and Brown said that a woman might carry a load of fifty pounds or more from her garden, Brown, 1978, op. cit., p. 44.

[91] Kenneth E. Read, The High Valley, London, George Allen and Unwin, 1966, p. 40.

[92] Ibid., p. 181; Brown, 1978, op. cit., pp. 44-45; and Brown and Georgeda Buchbinder (eds), Man and Woman in the New Guinea Highlands, Washington, American Anthropological Association, 1976, p. 4.

[93] Reay, 1959, op. cit.

[94] Reay, ibid., p. 192; and Berndt, 1962, p. 171.

[95] Berndt, ibid., pp. 167-168, 172-173; Shirley Lindenbaum, 'A Wife is the Hand of Man', in Brown and Buchbinder, op. cit., p. 56; and Reay, 1959, op. cit., p. 111.

[96] Reay, 1959, op. cit., p. 124; Lindenbaum, 1976, op. cit., p. 59; and Watson, 1967, op. cit., p. 270.

[97] Read, 1966, op. cit.

[98] Among the Gururumba, the first wife of 63 per cent of the men in Miruma and the second wife of 40 per cent ran away or was sent away by her husband 'because of her recalcitrance', Newman, 1965, p. 26; and among the Bena it was estimated that between 35 and 50 per cent of new brides attempted to escape at least once, Langness, 1969, op. cit., p. 45.

[99] Reay, 1959, op. cit., pp. 184-186.

[100] Read, 1966, op. cit.

[101] R.F. Salisbury, 'The Siane of the Eastern Highlands', in P. Lawrence and M.J. Meggitt (eds), Gods, Ghosts and Men in Melanesia, Melbourne, Oxford University Press, 1965, p. 67.

[102] Langness, 1974, op. cit.

[103] Langness, 1969, op. cit.

[104] Watson, 1967, op. cit.

[105] Nelson stressed the transformatory nature of the change among the Kaimbi, op. cit., pp. 214-215.

[106] Watson, 1979, op. cit., p. 67.

[107] Watson, 1967, op. cit., p. 245.

[108] Chinnery, op. cit., p. 118. Polygyny also existed of course in the non-sweet potato systems of Melanesia.

[109] Watson, 1979, op. cit., p. 67.
[110] Reay, 1959, op. cit., pp. 95 and 114.
[111] Read, 1966, op. cit., p. 65; Newman, 1965, op. cit., p. 59; Brown, The Chimbu: A Study of Change in the New Guinea Highlands, Cambridge, Mass., Schenkman, 1972, p. 111; Nelson, op. cit., p. 213; and Salisbury, 1964, op. cit., p. 227.
[112] Nelson, op. cit., pp. 213-214.
[113] Ibid.
[114] Watson, 1967, op. cit., p. 252 and 265.
[115] Langness, 1974, op. cit., p. 207.
[116] Ibid., pp. 193-194 and 207.
[117] Berndt 1962, op. cit., p. 171, suggested that men's antagonism for women was overcome only in sexual intercourse, but he showed, as did Reay and Lindenbaum, that intercourse itself incorporated coercion and violence. Coitus sometimes involved rape and this, outside the local group, was a form of political plunder. Lindenbaum, op. cit., p. 60.
[118] K.E. Read, 'Cultures of the Central Highlands, New Guinea', Southwestern Journal of Anthropology, 10, 1, 1954, p. 29; and Langness, 1974, op. cit., p. 202.
[119] Newman, 1965, op. cit., p. 41.
[120] Marilyn Strathern, Women in Between: Female Roles in a Male World: Mount Hagen, New Guinea, London and New York, Seminar Press, 1972, p. 47; and Newman, 1965, op. cit., p. 40.
[121] Strathern, ibid., p. 48.
[122] Salisbury, 1965, op. cit., p. 51; and Strathern, ibid., p. 13.
[123] Newman, 1965, op. cit., p. 42.
[124] Quoted in Lindenbaum, 1976, op. cit., p. 59.
[125] Reay, 1959, op. cit., p. 15; Read, 1966, op. cit., p. 141.
[126] From the account of Die Mbowamb presented by Andrew Strathern in The Rope of Moka, London, Cambridge University Press, 1971, pp. 204-208.
[127] Ibid., p. 205. Kuma big men also tried successfully to perpetuate themselves over time; Reay 1959, op. cit., p. 114, found in the 1950s that a leader was succeeded by his sons in 64 per cent of cases.
[128] Quotedin A. Strathern ibid., p. 205. Kakoli bachelors in the upper Kaugel Valley in the 1960s lived a life of social and physical abjectness and dependency, brought on by their lack of wealth, Nancy Bowers, 'Permanent Bachelorhood in the Upper Kaugel Valley of Highland New Guinea', Oceania, 36, 1, 1965.
[129] Waddell, op. cit., p. 217.
[130] Watson, 1979, op. cit., p. 68.
[131] As Robert Brenner said of class in relation to economic development in pre-capitalist and capitalist systems. 'The Origins of Capitalist Development: A Critique of Neo-Smithian Marxism', New Left Review, 104, 1977, p. 27.
[132] Watson, 1979, op. cit., p. 68.
[133] Watson, 'From Hunting to Horticulture in the New Guinea Highlands', in Langness and John C. Weschler (eds), Melanesia: Readings on a Culture Area, Scranton, Chandler Publishing Co., 1971, pp. 17 and 31.
[134] Thus Brown, writing as recently as 1978, said that salt from springs in the lower Wahgi Valley 'is distributed in trade throughout the Highlands' (our emphasis), and 'few workers migrate permanently and most Highlanders rarely leave their home regions for very long', pp. 41 and 130.

[135] Andrew Strathern, 'Gender, Ideology and Money in Mount Hagen', Man, 14, 1979, p. 534, and 'Finance and Production: Two Strategies in the New Guinea Highlands Exchange Systems', Oceania, 40, 1, 1969, p. 43. This was recorded when the production of coffee, in the hands of big peasants in the highlands was being organised on an increasingly large scale.
[136] Sillitoe, op. cit., p. 134.
[137] Ernest Brandewie, 'The Place of the Big Men in Traditional Hagen Society in the Central Highlands of New Guinea', Ethnology, 10, 1971, pp. 203, 205, 209.
[138] Sillitoe, op. cit., pp. 109 and 111.
[139] Emmanuel Terray, 'On Exploitation: Elements of an Autocritique', Critique of Anthropology, 4, 13 and 14, Summer 1979, p. 35, Claude Meillasoux, 'From Reproduction to Production', in Wolpe, op. cit., p. 195.
[140] Andrew Strathern, 'Descent and Alliance in the New Guinea Highlands: Some Problems of Comparison', Proceedings of the Royal Anthropological Institute for 1968, pp. 39 and 42.
[141] The men's house was of course the physical expression of this unity and separation. Further on big men and the men's house, see M. Donaldson, 'Contradiction, Mediation and Hegemony in Pre-capitalist New Guinea: Warfare, Production and Sexual Antagonism in the Eastern Highlands', pp. 435-459 in Melanesia: Beyond Diversity, Vol.2, R. May and H. Nelson (eds), Research School of Pacific Studies, Australian National University, Canberra, 1982.
[142] Terray, op. cit., p. 38.
[143] Watson, ibid., and Francis West, 'An Australian Moving Frontier in New Guinea', in Neil Gunson (ed.), The Changing Pacific: Essays in Honour of H.E. Maude, Melbourne, Oxford University Press, 1978, pp. 216 and 222.
[144] Talal Asad, 'Introduction', in Asad (ed.), Anthropology and the Colonial Encounter, London, Ithaca Press, 1973, p. 9.
[145] West, op. cit., p. 222.
[146] Bronislaw Malinowski, 'The Rationalization of Anthropology and Administration', Africa, 3, 4, 1930, pp. 419-429.
[147] E.W.P. Chinnery, 'Presidential Address: Applied Anthropology in New Guinea', Australia and New Zealand Association for the Advancement of Science, Report of the Meeting, 21, 1932, p. 166.
[148] M. Fortes, 'Preface', in Fortes (ed.), Social Structure: Studies Presented to A.R. Radcliffe-Brown, Oxford, Oxford University Press, 1949. Chinnery dated the introduction of the new courses (and the installation of Radcliffe-Brown) from 1924, but Fortes said 1926.
[149] West, op. cit., p. 222.
[150] Fortes, op. cit.
[151] Quoted approvingly in Chinnery, 1932, op. cit., p. 175.
[152] William C. Groves, 'Anthropology and Native Administration in New Guinea', Oceania, 6, 1, 1935, pp. 94 and 101-102.
[153] Chinnery, 1932, op. cit., pp. 163-164.
[154] Ibid., p. 167.
[155] Ibid., p. 171.
[156] Ibid., p. 164.
[157] West, op. cit., pp. 222-223.
[158] Williams obtained a Diploma in Anthropology at Oxford in 1922, became Assistant Anthropologist, Papua, in the same year and was Government Anthropologist, 1928-43; in 1922-1934 he studied under

Malinowski at the London School of Economics on a Rockefeller Fellowship. Francis Edgar Williams, 'The Vailala Madness' and Other Essays, edited with an introduction by Erik Schwimmer, St. Lucia, Queensland, Queensland University Press, 1976, pp. 11-13 and 419-421. Williams could however indulge himself in socially disruptive activity. In the mid-1930s he threw pearl-shell in handfuls from a reconnaissance aircraft flying over the Southern Highlands, an incident which must have had considerable impact on the social relations of people living over a wide area. Sillitoe, op.cit., p. 17.

[159] Elkin, 'F.E. Williams - Government Anthropologist, Papua', Oceania, 24, 2, 1943, p. 89.

[160] Asad, op. cit., pp. 11 and 17, and David Goddard, 'Anthropology: The Limits of Functionalism', in Robin Blackburn (ed.), Ideology in Social Science, Bungay, Suffolk, Fontana, 1972, pp. 61-62. Newman, op. cit., pp. 9 and 10, described in the 1960s how village leaders were summoned by the colonial officials to meet the anthropologist. He was saluted by them, and conducted to the men's house, where he stated his business.

[161] Hasluck, 'A Policy for New Guinea', South Pacific, 5, 2, 1952, p. 227.

[162] Waddell, op. cit., p. 219; and Brown, 1978, op. cit., pp. 103-112. The Simbu Valley, West said, 'was probably overpopulated in pre-European times'. F.J. West, 'Indigenous Labour in Papua-New Guinea', International Labour Review, 77, 2, 1958, p. 104.

[163] Mervyn Meggitt, Blood is Their Argument, Palo Alto, California, Mayfield Publishing Co., 1977, p. 182.

[164] Read, 1954, op. cit., p. 10, and 1966, op. cit., p. 141.

[165] Yen, op. cit., p. 116.

[166] Brookfield, 1964, op. cit., p. 38.

[167] Read, 1954, op. cit., 'even before the coming of Europeans, land was sometimes sold'.

[168] Robin Radford, 'Missionaries, Miners and Administrators in the Eastern Highlands', Papua and New Guinea Society Journal, 6, 172; and West, 1978, op. cit.

[169] Reay, 'Generating Political Conflict: Some Consequences of Economic Exploitation of the New Guinea Highlands', Anthropological Forum, 3, 1973-4, p. 297.

[170] Uneven development, in favour of both the highlands and islands, characterised capitalist agricultural development in PNG generally.

[171] Richard Salisbury, 'Early Stages of Economic Development in New Guinea', Journal of the Polynesian Society, 71, 1962.

3 The politics of colonisation

Foreigners seized power in New Guinea with, so colonial history reads, a minimum of violence and maintained it with less. The distortion in this view arises precisely because it is a colonial view which neglects the political and economic context into which the Europeans intruded. It begs the question of who fought whom, who died, and who held power.

The most famous highlands explorer and prospector, Michael Leahy, himself remarked, 'New Guinea had been explored and pacified only by bloodshed', [1] and there can be no doubt that the number of Papua New Guineans who lost their lives is greater than commonly supposed if only because the number killed in the Mandated Territory of New Guinea was not reported until after the Second World War. [2] Of the white casualties there is greater certainty. During the first quarter century of colonisation almost every German factory and plantation experienced bloody incidents, and 55 out of the 230 enumerated white population were killed in that period. What is much less clear is the size of the reprisals meted out to the Papua New Guineans for what Reed called 'armed uprisings'. 'Large numbers' of villagers were slaughtered in a punitive expedition following the killing of 4 Polynesian Wesleyan missionaries around 1880 in Blanche Bay. [3] Similarly, Leahy commented ominously after the punitive action following the death of the miner McGrath in 1933, 'I think we gave Ekanofe a fair idea of what a crime it is to kill a white man'. [4] Under German colonialism near the turn of the century, the death of two whites and eight indigenous workers in the St. Matthais group led to the execution of more than eighty villagers, and the death of one white man in the Huon Gulf brought the death of forty. [5]

Leahy recorded occasions in which patrols wounded attackers and drove them off, or were forced to fire blindly into bush or kunai. [6] The numbers killed in beating off unsuccessful attacks were seldom recorded, and where they exist such estimates must be conservative. This is part of the context within which the 'opening up' of the highlands by Australian colonialism must be placed. Ian Downs laconically remarked that 'fatal patrol officer casualties averaged a little over one a year up to 1939', [7] a high proportion given the very small number of kiaps actually actively engaged in advancing the 'moving frontier'.

The deaths of whites engaged in the penetration of the highlands must be balanced with the reports of later social scientists such as Sorenson that 'immense popularity was the lot of the early patrol officers'. [8] The number of published accounts of attacks on colonial patrols is large. It includes, for example, the two attacks on the Leahy-Dwyer patrol of April-May 1931 which resulted in the wounding of the two Leahys, and the attack on 20 November 1932 on Leahy's patrol prospecting on behalf of New Guinea Goldfields Ltd. Mick and Dan Leahy later helped the kiap 'round up the murderers' of the miner McGrath and in the process were themselves attacked yet again in April 1934. [9] These attacks were preceded early in 1931 by a well-planned ambush of the Leahys and Dwyer prospecting patrol in 'Kukukuku country' which took place at the same time as the miner Baum was killed. Of the patrol, four were wounded, and 'at least six' of the attackers were killed. [10]

The missionaries were attacked as well. The first foreigners to enter the highlands in 1919 and 1920 received a hostile treatment in Gadsup and withdrew. In 1934, a mission house was burnt to the ground in Simbu, and in the east, native evangelists were involved in clashes in May 1936. [11] Patrol officers came under more frequent attack. Taylor was forced to shoot in the Siane and Sinesine areas in August 1933. In August 1938, Medical Assistant Walsh and 7 native police were attacked by 6,000 warriors at Wabag. Two police died, and Taylor, Downs, Edwards and 15 police reinforcements, came in by air to relieve the beleaguered garrison. Downs was attacked three times in the Gai Valley near Hagen shortly afterwards while repatriating carriers from the Taylor-Black patrol. [12]

The attacks by highlands warriors on those from outside the region led to a tightening up of the Uncontrolled Areas Ordinance. In 1935 the tougher regulations prohibited new entry into or exit from the highlands for non-government people. Those already in the highlands were permitted to stay, but restricted to their camps or stations. [13] The regulations remained in force until January 1936, when the 1925 Ordinances were reintroduced with the additional provision that travel was restricted to a five mile radius of the camp or station. Missionaries were permitted to continue to visit only those villages approved by the kiap and marked on a special permit. [14] In October 1939 acting under the ordinances, the Director of District Services and Native Affairs recommended that 50 per cent of the native evangelists at the Ogelbeng mission station be removed. [15]

The attacks on the white vanguard led to a militarisation of exploratory patrols. Leahy and Taylor were accompanied on their 1933 patrol by 20 armed carriers and 12 police [16] and of his final major patrol in 1934, Leahy commented favourably on the 'increased discipline and efficiency of our little army' which comprised 25 armed 'boys' who carried no baggage at all. [17] Even with such a force Leahy was ambushed several times, his men killing at least three warriors. [18] Something of the desperation of the situation can be gauged by the fact that at the closure of the Hagen-Sepik patrol in 1938, colonial officials withdrew to 'a line drawn 10 miles west of Hagen'. [19]

Over an area that later comprised three sub-districts of the Eastern Highlands, patrol officer Atchison supervised the Kainantu mining operations, enforced the labour regulations, surveyed and purchased land for the Aiyura agricultural station and the town of Kainantu, and undertook agricultural experiments. [20] Others were equally hard pressed. Shortly following his air-rescue of Walsh, Downs was appointed to Simbu. Years later he reflected that 'large dynamic groups increasingly became convinced that we did not have the ability to enforce the law and order that we preached'.

Yet the spirit of colonial adventure among the kiaps was strong, as was the notion of their being 'set apart', even 'specially chosen'. Their isolation and immense power encouraged such pretensions. As Downs admitted,

> I like to think that George Greathead and I were still conscious of the fact that we had too much power for young men remote from real authority. I know that I was glad to escape from so much blind submission when I finally got away to the war. [21]

Radcliffe-Brown, as noted, took the founding chair of anthropology at the University of Sydney and assisted in establishing a programme for cadet patrol officers. He taught the anthropology course himself and was followed by one of his first pupils, Hogbin, and in the chair by Elkin. [22] 'They would have been surprised to know', said Downs, 'the lasting value of their efforts ...' [23] They taught the axiom of Radcliffe-Brown that 'any persisting culture [is] an integrated unity or system in which each element has a definite function in relation to the whole', [24] and was interfered with at peril.

In 1936, when Downs arrived in New Guinea, the Department of District Services and Native Affairs was staffed by fewer than 100 officers who controlled an enumerated population of about 500,000. Three years later, the kiaps with only 15 patrol posts and 934 native police, controlled an enumerated population of about 600,000. [25] In the field, the situation was even more acute, and in Simbu, Downs and 17 native police confronted 300,000 people. The functionalism of Radcliffe-Brown, Hogbin and Elkin had made a virtue out of necessity.

The colonial pathfinders

To enter the highlands it was necessary to climb more than 1,500 metres from the Markham Valley before crossing the mountain wall into the high valley systems, themselves separated by steep divides. Luxury goods, various commodities, and people moved along these routes to and from the lowlands.

Lutheran missionaries followed the practice of stationing native evangelists at points along the borders of uncontrolled territory. War allies and trading partners of villagers among whom an evangelist dwelt were usually the first to be contacted when the missions pushed into new territory.

In 1919 the Lutherans at Kaiapit were invited by evangelists working among the Atzera in the Upper Markham Valley to visit the Atzera's trading partners in Gadsup in the Eastern Highlands. [26] Toward the end of 1919 the missionary Stephen Lehner accompanied by Atzera guides crossed the ranges from Kaiapit to Bunanamu village in what was later the Gadsup sub-district, but the group returned after one night there. They were followed in February 1920, by the missionaries Pilhofer, Oetel and Stoessel, the evangelist Gape'nuo'ngizaki, 25 men from the Sattleburg congregation which was to provide the native evangelists, and guides from the Atzera. The party progressed slightly further than Lehner, but came under attack near Pundibassa. Gape'nuo' and two white missionaries were wounded and Makakajo, the big man of the Pundibassa, was killed and several of his warriors wounded. [27] The party withdrew, but a few years later the Sattleburg congregation sent its evangelists into the Eastern Highlands to stay. The missionary Flierl visited the evangelists annually and between 1926-1928 made three exploratory journeys through the district. [28]

The major breakthrough into the centre of the Eastern Highlands followed the 'discovery' by Gape'nuo' of a trade route leading from the middle Ramu Valley to the village of Lihona (Rihona; Lehuna) on the Dunantina headwaters of the Purari River system. [29] Here the Lutherans established 'a small community of native converts to learn the language of the mountain tribes', and from here Dwyer and Leahy set off on their 1930 expedition. [30] Nor were missionaries alone in their use of long existing trade routes as paths into and through the highlands. The missionary Ross stated that Leahy and Taylor followed a 'broad trade route, probably the main Papuan coastal trading route inland'. [31]

Although the reaction to foreign penetration was often hostile, the colonisation of the highlands would have been immeasurably more difficult if it were not for the co-operative attitude of some highlands big men. Perhaps it was the case that the enemies of those who attacked the intruders moved to consolidate their own position by taking the opportunistic decision to treat the newcomers as allies. Certainly it was the big man Rujao of Lihona village who made Gape'nuo' so welcome and Kavagl (sometimes Kauwagl), the Simbu big man of Korugu, under whose 'direction and guidance' three catholic missionaries crossed the Bismarks in November 1933. [32] Similarly, it was Nindi the big man of the Mogei-Nampoga clan who delivered to Fr. Ross the first batch of 50 boys aged 10-16 for the catholic mission school in Hagen in March 1935. Of these, 28 were baptised in December 1938, the first catholic Hageners. [33]

The status of other known collaborators is not so clear, but their role was nonetheless crucial. When Taylor was attacked in the Siane in 1933, peace was maintained largely by Heruku from the Yamofwe people who, given his ability to mediate in a conflict situation, was presumably in a leadership position. Another man from the Yamofwe, Lonbonguie, 'a man of some local renown', acted as a guide, interpreter and sponsor, and took a leading part in introducing the party to the people of each new village, persuading them to supply food and then acting as a middle man in trading transactions. [34]

Trade routes in the New Guinea highlands involved complex political and social matters, the final adjudication over which was predominantly in the hands of big men. Foreigners, like trade goods, were passed from hand to hand along a complex chain of alliances, [35] a product of constant warfare and subject to rapid change and realignment. Incessant warfare prevented a united resistance and necessitated the complex and shifting alliances utilised by the colonial pathfinders.

The role of the police

Along with the multitude of carriers who accompanied the early patrols, came also the members of the Native Constabulary. It was upon these men, probably more than any others, that the work of pacification devolved. Certainly the Treaty of Capitulation between Germany and Australia recognised the need for their continuation, and they remained under the Australians the direct instrument of colonial state power, the 'visible cutting edge of the Territorial Government'. [36]

By 1938 the number of native police stood at nearly 900, but on patrol, at one particular place and time, they would number seldom more than a dozen. [37] Following the work of the initial penetrative patrols, camps of native police were frequently placed at strategic points in the area, sometimes left alone, and often, given the high rate of transfer among <u>kiaps</u>, left to their own devices for long periods. [38] Picked native constables were sometimes sent out with their wives and families to live in key villages. [39] Downs has noted that the history of base camp police was that they either 'abused their authority or lost control by default to astute local leaders'. [40]

The constables survived, and indeed flourished in the hostile environment, not simply because they enjoyed the advantage of the rifle against the unfletched arrow, but also because the role they played was already well established, prestigious and powerful in a highlands polity. Police constables were essentially warriors, 'savages in serge', the Australian Administrator, Hubert Murray, called them, [41] and in the recruitment of constables the administration preferred raw recruits, not men who had been mission-trained, or were finish-time labourers. [42]

The behaviour of the native police was redolent of the behaviour of warriors, as the <u>kiaps</u>' was of big men. The <u>kiaps</u> killed pigs, controlled trade, administered justice and enjoyed the obedience of their warriors. They also exhibited the personal traits of arrogance, aggression and authoritarianism which were the hallmarks of the greatest of big men. It seems that the <u>kiaps</u> not only acted, but were perceived, as big men, and the police as warriors and/or big men depending on the proximity of white authority.

One of the first actions of Downs in Simbu was to enter an alliance with the leaders of Siambuga Waugwa and the Kamanegu, and to take the field with 2,000 warriors on his side against a particularly troublesome warring big man. In forming the alliance, Downs was acting in the tactically approved manner, one well understood by his big men allies and foes alike. [43]

Less constrained by colonial legalities, the native constables saw little need to restrain themselves in the role of fight leader. In 1946, police united with Kamano warriors whose enemies they attacked in response to a bribe, an event unremarkable in politics noted for shifting alliances and payment for military assistance. [44] Indeed, Berndt states that police away from the main station 'encourage[d] the practice of bribery which [was] after all, traditionally based ...' and that 'it [had] much the same function ... as in the past in efforts to obtain allies in interdistrict fighting'. [45]

Police behaviour toward women who broke the law, or to the wives of men who had offended the law, predictably was exactly what was expected of victorious warriors. [46] Their behaviour toward women in general was reminiscent of that of the great warrior leaders like Matoto of Tairora (see chapter 2).

> Native police of this region commonly avail[ed] themselves of local women; they [did] not keep permanent mistresses but prefer[ed] variety. This [took] place whether or not their own wives [were] present. The guardian or husband of the women abducted by a native police officer rarely receiv[ed] compensation, nor ha[d] he any redress; he [was] usually afraid to approach the administration officials at Kainantu or elsewhere because of the power of the native police. [47]

Alpers and Gadjusek noted that following the establishment of a police post at Okapa the surrounding villages became infected with gonorrhea which spread in 'epidemic fashion' through the central North Fore. [48]

To end warfare and cannibalism, and for the pacification of an area, the kiap needed an efficient, effective police force with a good grasp of local conditions. Indigenous warriors, far from home, armed with rifle and bayonet and backed by the majesty of the colonial state, filled the role handsomely.

Colonisation and missionaries

The Rev. Dr. Christian Keysser, a missionary among the Kate above Finschafen, 'developed a system whereby Christian 'helpers' or colonists ... would settle in a pagan area and give living demonstration of the Christian life'. [49] Another white missionary explained that this method of proselytising was required because 'all the attempts of the mission failed at first because of the very close solidarity of the natives'. [50] Since it was not possible to convert individual men and women, the Lutherans aimed to convert entire villages through their 'colonisation' by 'helpers'. Christian evangelists, who settled in a village to demonstrate the efficacy of the new faith, 'waited patiently until the whole tribe was ready to accept Christianity'. [51]

Sometimes that wait was long; it was 13 years before the Lutherans at Mt. Hagen obtained their first converts. [52] But the rewards for patience were great, for the newly converted community in turn provided evangelists to colonise other communities. It was the rule that from each major group of new converts some would volunteer for the work of evangelists [53] and indeed one missionary explained that 'missionary service [became] so natural that it [was] a disgrace if a congregation [had] no evangelist of its own among the pagans'. [54] The self-perpetuating nature of this form of missionising was even more useful because its implementation was extremely economical. Evangelists and teachers were supported by their home community, until they had effected the conversion of those among whom they dwelt whence they were supported by them. The new christian community in turn sent out and supported its own evangelists in other communities. Just as the colonised met part of the cost of their colonisation through head tax, so too did the evangelist system ensure that they paid for their own conversion.

The mission stations, along with the gold miners, were the first foreign economic enterprises in the highlands to rely on hired labour. The first station in the Western Highlands, built by '15 coastal boys and hundreds of local helpers', was very large, and it was maintained by '15 coastals and a labour line of 20'. [55] At Anggil in the Ulga Valley, in one day an outstation of 11 houses was built with the help of 2,000 people. [56] In 1935 the mission purchased 200 acres at Rebiamul near Wilya and set about building permanent wooden buildings. The preparation of the timber alone for the new structures involved the labour of 48 men. [57]

A missionary with the Christian Missions in Many Lands organisation, O.C. Fountain, commented that mission stations were the 'religious equivalent of plantations in the economic realm'. [58] A mission station could easily have a population of 200, comprising pupils, workers and evangelists, and their families. Fountain calculated that the Virnum mission station in Sepik where he worked, absorbed the surplus of no less than seven surrounding villages. [59]

Competition between the Lutherans and the Catholics, the two major missions in the highlands, was fairly intense. Native catechists and teachers were placed in as many highlands villages as possible 'to peg out a spiritual claim'. [60] By 1932, from Bena east to the Morobe border, eight village mission posts were in operation. Two years later the number had more than doubled to 19 and each was staffed by one or more evangelists and their families. [61] By 1935 there were 10 white missionaries and 74 helpers in the highlands. [62] About 3,000 evangelists, three for every one native constable, were spread throughout New Guinea by 1938. [63]

It was the application of christian teachings to the world of the highlands village that proved difficult. The evangelist living in the village was dependent upon it for land and protection. He was recruited to the village in a manner similar to the recruitment of refugees and others, and was thus at the mercy of the big man and his followers. The problem confronting the evangelist was to survive by remaining a client of the big man, but at the same time to undermine the ideological structure which supported him.

The missionaries clearly perceived what Aufenanger referred to as the 'War-Magic Houses' [64] to be a main target in their task of ideological subversion. The missionaries were, according to Berndt, at least nominally committed to 'undermining almost all traditional belief and activity as uncongenial to christianity or to their version of it.' [65] Reay recorded the following oration from the Minj-Wahgi:

> The kiaps came up and our traditions went down.
> The missionaries came and set fire to our war-magic houses.
> Kiaps and missionaries, missionaries and kiaps, these white men
> Have thrown away our ancestors and put us in pain. [66]

In the Sepik in the 1920's a missionary dragged four women through the men's house in an attempt to break its power, just as in New Ireland in the 1880's missionaries raided sacred sites, stealing and destroying sacred symbols. [67] No new men's houses were built by the Gehamo-Nagamidzuh once the village decided for Christ in 1950, and the rituals associated with nama and male initiation were curtailed. [68] More slowly, but just as definitely, the same thing occurred among the Bena and the South Fore. [69] The colonial state, too, considered the men's houses 'a definite military objective' [70] and by the late 1950s, the combined efforts of the missions and the government had reduced the number of men's houses to 'only a few' in the Wahgi Valley and adjacent areas. [71]

Combined with the attack on the old institutions and rituals came an insistence upon an end to polygyny. [72] The missionaries refused baptism to a man with more than one wife and thus, for example, the baptism of the Gehamo-Nagamidzuha was preceded by a number of divorces. [73] Reed concluded, however, that these 'direct attacks on the native folkways of marriage ... were unavailing'. [74] With pacification the opportunities for, and the tendency toward polygyny, further accelerated. The missions taught against what remained and was increasingly an important economic institution. The reasons for the failure of the missions to eradicate polygyny, at least among those with power, the big men, lay in the vital role women played in the village economy. The big man was a polygynist for reasons that went to the heart of the system of agricultural production in the region.

The combined government and mission thrust to eradicate the men's houses was more successful, because the external role they played was removed with conquest and pacification. The big man institution survived the missions and pacification, and perhaps even flourished with the latter, a state of affairs incongruent with the missionaries' other gift, monogamy. A big man could not survive the end of polygyny. Even if the other members of his village were to embrace monogamy, his own polygynous position was intimately and directly linked with his ability to assume his role of village intermediator. Monogamy would weaken a big man's ability to relate as a big man to other villages. More, since it was at the root of his own personal wealth, it would mean a decline in power within his own community. For the monogamous villager, for the rubbishman, the church made a virtue out of what was economic necessity. Christianity was indeed a religion for the meek and humble. For the big man, the teaching on monogamy was just one more convention to be flouted, one more rule, the breaking of which would enhance not diminish, his authority.

It is scarcely surprising then, as Vicedom pointed out, that 'God did not choose the men who were also leaders among the people'. [75] Nor should the missionary Kuder have found it strange that the church 'lack[ed] outstanding men of ability to represent it...' or that 'ambitious men [might] feel frustrated in Christian work and [seek] advancement in secular occupations'. [76] The big man, Bumbu of Busama, was twice condemned by his congregation, and once suspended from membership of his church. It made no difference to him. [77] The leaders of the people, the ambitious men of outstanding ability, did not countenance subservience to the will of a native evangelist/refugee, but they very quickly found ways in which missionaries and other white invaders would be useful to them.

Big men at the beginning of the transition

By the 1920s at least, new plants and crops were filtering into the highlands through trading networks. Maize, new varieties of sweet potato, and the now ubiquitous casurina tree, were gradually adopted. Steel axes, cloth, beads and other goods also reached the people of the highlands, perhaps even before the existence of highland groups was suspected by the whites on the coast. [78] Leahy reported sighting two axes and the steel handle of a knife on his first patrol in 1930, and on his 1934 expedition he reported 'a large number of steel axes and one big knife'. [79]

The introduction of steel axes reduced the workload of men by some 30 per cent which led in turn during the period of 'indirect contact' to a 'florescence of ceremonial and warlike activity'. [80] With men spending less time on gardening activities and more in directly political activities the demand for pigs rose as the number of ceremonial participants increased, and the number and size of ceremonial payments roughly trebled. [81] As pig production rose (by an estimated 30 per cent in Siane) so did pig stealing and the disputation due to it. [82] Both increased fighting and payment inflation effectively placed even more power in the hands of the big men. [83]

The value of women and pigs rose relative to shells and stone axes, and in Siane, bride wealth payments had doubled by 1940. The effect of this inflation was to increase the power of big men relative to that of the 'nothing men' [84] and to accelerate the centralisation of wealth between communities. The centralisation and concentration of wealth and power was further intensified as whites followed their trade goods into the region.

The establishment of mission stations and patrol posts in the vicinity of a particular village or group of villages accelerated the concentration of wealth, as did the presence of even passing patrols. Whenever a patrol arrived, goods were given to the big men who represented their villages and made first contact. The big men likewise hastened to give presents of pigs and to receive presents of shell in return. [85] Ross commented in 1935, 'The coming of Europeans to this area [Hagen] has not materially affected native customs of trade, except to increase tremendously the wealth of natives close to European camps'. [86] Some astute leaders realised quickly the advantage of having a resident white of whatever capacity. When the anthropologist Newman arrived to live with the Gururumba, 20 miles from Goroka, he was told, 'You have come now everyone [other native groups] will eat our pigs'. [87] The Gehamo and Nagamidzuha of Goroka sold land just to obtain their own white [88] and the Fore actively competed for missionaries. [89]

The wealth of a big man and the size of the village were closely related, the former drawing in followers and other newcomers and the greater numerical strength in turn making the village more secure militarily. While on patrol with Leahy, the kiap Taylor observed that the villages in a particular area tended to be of very nearly the same size. Taylor and Leahy agreed that if a village became weaker than its neighbours, it would 'sooner or later be wiped out'. If, on the other hand, it became very strong, 'a coalition of neighbouring villages [would] likewise wipe it out, burning the houses, destroying the gardens and ringbarking the trees'. [90]

Colonialism upset this balance of terror by affording some villages a rapid increase in wealth. The result was an escalation in warfare. Leahy noted on returning to Bena:

> We were appalled at the number of violent deaths in the neighbourhood during the few months we had been away. Each day we heard of more killings, most of them in war... [91]

The increasing concentration of wealth, coupled with an intensification of fighting, led in turn to the development of larger defensive units and so to 'an increase in the strength and authoritarianism of traditional groupings'. [92]

If wealth and military power were becoming increasingly concentrated, what were the political and economic gains, so clearly recognised by the big men, that pacification brought? Downs commented that 'indirect rule and the operation of the common law began with tremendous co-operation from less than a dozen really influential fighting leaders'. He added, 'They were not just helping us. They were helping to maintain their own economic and social position'. [93] Reay noted of the Kuman big man Konangil, that his 'clan had already achieved remarkable success in warfare, and he was quick to realise that a ban on further wars would consolidate its position'. [94] Pax Australiana froze the political status quo. When the police bois marched in and peace broke out, what was, continued to be with no fear of reversal. Those who had won through war, won even more through peace.

The 'considerable political consolidation' remarked by Salisbury [95] that occurred during the first phase of colonial 'contact' was accelerated by the appearance of government patrols, one of the most important tasks of which was the taking of the census. The patrol officers assigned village names somewhat arbitrarily as they tried to establish census units. All people were assigned to one or other of these census units, usually called 'villages', and were required to be physically present at their village for the biannual census. [96] The effect of the census in some areas was to 'make the people feel harnessed to the place where their name [was] recorded in the village register'. [97]

The colonial patrols brought with them not only a 'book of names' but also new physical structures which became a focus for resettlement. Thus, shortly after the decision to establish a patrol post at Lufa, patrol officers began to build rest houses, and missionaries began to construct similar buildings for the holding of services. The Fore shifted their own dwellings nearer the white presence to take advantage of the goods and knowledge that came with it, and in so doing created much larger villages. [98] The new villages, for example among the Gururumba, were as much as 3 to 6 times larger than those built before colonial intrusion, [99] and among the Siane the size of the 'effective political unit' rose significantly from about 80 individuals in 1933 to 200 in 1952. [100]

The new peace assisted in the expansion and consolidation of trading and ceremonial linkages between political groupings, reinforcing the centralisation of women and the concentration of wealth. Peace meant that such networks could grow larger and the pool of women available increased for those with the wealth to acquire them. [101]

In 1953, Patrol Officer Vial noted a shortage of women and pigs in some areas and their tendency to flow toward those groups wealthier in axes and shell. [102] As the wealth of individual big men increased through inflation and 'boom', so did the possibilities of, and the necessity for, this wealth to be invested in the acquisition of women. When the 'castle walls', the stockades and palisades which surrounded the village were pulled down at the orders of the patrols, the big man's warrior role fell with them, but his entrepreneurial, ambassadorial and labour management roles developed further.

One of the earliest effects of the introduction of steel implements, which found their way into the hands of the wealthy, was to increase the demands on, and the demand for, women agricultural workers. The blossoming in trade and ceremony was built on an intensification of labour, for larger pig herds required increased agricultural production. The reduction in labour time possible through the new technology was lost to women in the cultivation of larger areas of sweet potato, an event caused, as Salisbury pointed out, not through the spontaneous action of women, but by the conscious decision of men. [103] The labour time required for the fencing of the greater acreage by men, did not increase as sharply as the labour time commitment of the female cultivators, for the key to maintaining larger pig herds was to abandon the gardens to them earlier for forage, before the fence repair work grew too sharply. In this manner, an area 40 per cent larger than previously cultivated, required only 12 per cent increase in the labour of men, whose labour input had anyway been reduced by 30 percent with the introduction of steel. [104] Already by 1953 in Goroka, one man who 'paid' his nine wives had become privately wealthy, even by European standards. [105]

Luluais and tultuls

The appointment of <u>tultuls</u> and <u>luluais</u> was one of the principal means of bringing a village under governmental supervision and control. The <u>luluai</u> was responsible for order in the community under his supervision. He had to prevent his group from warring with others, to carry out directives given him by the government and bring to the attention of the patrol officer any matters of dispute that could not be settled in the community. He also kept possession of the village census book. The <u>tultul</u> was an assistant to the <u>luluai</u>, and sometimes possessed <u>Tok Pisin</u> skills that the <u>luluai</u> lacked.

The first luluais in the Eastern Highlands were appointed in villages close to where the kiap Ian Mack was killed, with the obvious object of preventing further losses. By 1939, Ian Downs had appointed 300 bosbois (headmen) in the Simbu Valley, 20 of whom he recommended for appointment as luluais. [106]

'The qualities which made for successful leadership under the traditional system ... seem[ed] to be no less important in the new social configuration', [107] and so the village officials were selectively chosen [108] because of their power prior to colonial intrusion. As Alpers and Gadjusek explained the situation, 'early Administration officers were careful to choose for their appointed headmen in each village those who carried the most influence, and, in general, they chose well'. [109] Downs' criteria were that the leaders selected 'could dominate other headmen in the clan'. In Simbu most of the 300 bosbois appointed by him were former fight leaders. [110] The kiaps who made the first appointments tried to ascertain which men already possessed power which the government could use to further its aims, and most of the first luluais were men who had made their names as warriors and persons of substance in their communities. [111] 'Government policy [was] to appoint the man who seem[ed] to be the most outstanding personality as the village chief with the title of luluai'. [112] The first luluai appointed among the Kuma was Konangil under whose leadership the 'strong and numerous Konumbuga routed their enemies and scattered them far and wide'. [113]

A big man whose career stretched 20 years like Kongangil's and Matoto's was rare in pre-colonial New Guinea. With social manhood not achieved until a warrior's late twenties and a life expectancy of less than 50, that the position could be held that long at all, was indeed remarkable. With pacification the big man's greatest fear and most frequent demise, death in warfare or through assassination, was considerably reduced. Big men were frequent targets of ambush and attacks by hired assassins, and with pacification came an end to some of these fears, particularly for the village officials, whose attackers could expect swift retribution from the native police and kiaps. Village officials were the 'last link between Central Government and the people' and were afforded close support by the patrol officers. [114] Government office then, made a big man's position more secure, and perhaps 'guaranteed a man's authority to an age when, in pre-colonial days, he would have slipped into the background', [115] or have been slain.

There are some differences of opinion, however, over whether official appointment meant an increase in or lessening of the big man's power. The variables involved were many, including the extent of the big man's power and influence prior to his appointment, his day to day relationships with the native constabulary, evangelists, teachers and other appointed officials. Also important were the closeness of the village to the government centre, the frequency of patrols, and the rate of turnover of colonial staff. When these factors were present in a favourable configuration, the result was often the emergence of 'small 'empires' ruled by dictatorial big men'. [116]

So embedded was the big man institution in the political economy of the highlands, that when appointments were made which ignored its efficacy, the big men, not the luluais and tultuls held power. Thus, in Korofeigu where according to Langness 'it was obvious that those men who were luluais and tultuls were not leaders under the former conditions', the real power in the area remained with the big men. [117] Berndt notes similar occurrences in Fore, and Hogbin warned that the 'system worked reasonably well as long as the luluai [had] risen to a position of supremacy by traditional means'. [118]

The durability of the big man institution is clear also in that those appointed by colonialism acted, in Barnes' words, 'as if they were chiefs'. [119] Similarly, some big men were able to appoint themselves luluias and tultuls, in that by acting as such officials de facto, they became them de jure, even to the point of replacing formerly appointed officials who lacked sufficient countervailing authority.

Reinforcing the colonial grid

The colonialism Australia imposed upon New Guinea resulted in the creation of a fine mesh of indirect control comprising in 1938, 11,608 village officials in 4,000 communities, nearly 3 for every community; plus 3,000 mission teachers, and about 900 native constables. [120] By 1962, the Eastern Highlands alone contained 1,900 luluais and tultuls, perhaps 1 for every 100 villagers. [121]

Downs was convinced of the efficiency of the colonial grid in that, in an unspecified period, 85 per cent of the information laid in charges before the Court of Native Affairs came from headmen, and 90 per cent of the prisoners were brought in by them. [122] The missionaries played their part in supporting lawful authority also, in that they were required to submit an oral monthly report to the patrol officer, as well as a written diary of their activities and impressions of the local situation. [123] The evangelists in the Kainantu area were commended by patrol officer Elliot who wrote of them in 1937, 'they have given every assistance to the patrol ... they have been of assistance in consolidating the work done by it'. [124] Berndt noted that they 'reinforce[d] and underline[d] the injunctions of the native police and administration'. [125] In most villages the mission evangelist kept simple records of births, deaths and marriages. These were, according to one veteran kiap, 'a great assistance to the patrolling officer'. [126]

Although the evangelists attempted to place themselves on an equal footing with the native officials, they were handicapped by the hostility of some kiaps toward them and by their lack of 'traditional' power. Writing at the end of the 1930s, Reed noted that the 'hat men' in some districts thought of themselves as a group apart from ordinary villagers, and that among them, the police boi held the greatest prestige. [127] Berndt was of the opinion that they had 'virtual control over all the officially appointed dignatories in their region'. [128]

The ability of the kiaps to exercise more direct policing functions, for example in the putting down of outbreaks of violence, was related directly to the establishment of an efficient road transport system, particularly in the Eastern Highlands. [129] As Bouchard has illustrated, the radial organisation of roads in the early period of white intrusion was related to the needs of administration, [130] leading to the establishment of what Ward and others described as a dendritic road network. [131]

One of the major tasks of the patrol officers was the building of roads to link the patrol stations. In 1952, in under six months, kiaps supervised the building of a vehicular road from Goroka to Kainantu to Gusap. Some labour gangs on the road construction comprised as many as 9,000 workers. [132] Another early priority was the building of the road linking the government station at Okapa with the sub-district headquarters at Kainantu. [133] Similarly, the first task of the patrol officer at Lufa was to link that station with the Highlands Highway. [134] In this respect, an important task of the luluais was to organise the village to undertake one day's labour each week on public works, usually road maintenance. The Lufa road was tackled with such enthusiasm that 'an occasional food shortage was created when some men neglected their gardens'. [135]

The construction of roads in turn reinforced the centralising tendency referred to earlier. 'Hamlets moved down from their ridges in order to be nearer the road, frequently consolidating with others as they did so' and the larger roadside villages in turn attracted desired visits from missionaries, traders and kiaps. [136] In Lufa, some villagers shifted their villages as much as 300 metres up the mountains to be closer to the road. [137] The increase in size and the centripetal relocation of villages, as noted, was linked to the centralisation and concentration of power within them. But while the key external contradiction which faced the big men, that of the problem of warring neighbours, was largely removed, the internal contradictions of sexual antagonism, remained and were probably heightened. The exigencies of the kiaps' situation required that as far as possible disputes within the village should be settled in accordance with custom. According to the missionary Nilles, kiaps were told that if such a settlement was not reached, indigenous justice would be carried out anyway. [138] Downs noted, 'When civil disputes and minor criminal charges were referred to the attention of the patrols, a village consensus might suggest what the best solution should be'. [139]

But of course the big men did not wait on the instructions of the kiap, or even his blind eye, to continue exercising power in the village. What Berndt termed 'embryonic courts' were in operation in parts of the Kainantu area even before those areas had been patrolled. [140] He suggested that their existence remained unknown to colonial officials for some time, even though their main instigators in the patrolled areas were village officials and native police. [141] Where they did come to the attention of the kiaps, Downs claimed that tacit official local approval for the existence of the de facto courts was sought and received. [142] In the other main area of commercial agriculture in Papua New Guinea, the Gazelle on New Britain, unofficial village councils had been operating at least since the mid-1930s. [143]

The informal courts arose directly from the need of those with power to have some instrument of domestic mediation for problems formerly resolved by the men's house. Their rise corresponded with the decline of that pre-colonial institution and the appointment of government officials. The informal courts were convened mainly by the native police, luluais and tultuls and occasionally by evangelists, either separately or in conjunction. Of 33 courts observed by Berndt 12 were convened by luluais and tultuls singly or in combination with each other; 9 by police singly or in conjunction with luluais and tultuls and 3 by luluais in conjunction with evangelists or 'mouth-boys' (unofficial officials); and the remainder were conducted by 'mouth-boys' and/or evangelists. [144]

Officially, the procedure was illegal, but as with the behaviour of the native constabulary, the kiaps were in no real position to prevent it. Reed had a more judgemental understanding of colonialism's relationship to the informal courts, suggesting that the majority of kiaps were 'lenient sometimes lax, in their attitude to illegal punishment. Usually it is only when natives have been seriously injured or have died as a result of rough treatment that the cases are brought to court'. [145] Certainly the punishments meted out to offenders were the same as those that would have been inflicted in the days of the men's house, with the exception of capital punishment, as were the charges laid and the cases heard. The 'continuity was located primarily in the content itself, especially in relation to procedure against the transgressor and the punishment meted out to him'. [146] A new institution, the informal court, enshrined the values of the pre-colonial men's house.

Both the punishments and the offences bore a striking resemblance to those handled prior to conquest by the men's house and as in the earlier period power and control were the essence of the procedure. As Berndt rather delicately put it, 'The informal court was not a court of equity'. [147] The treatment of accused men depended on their wealth and power, as is clear from the examples of a luluai who presided when he was the accused, and another who presided when he was the aggrieved party. [148]

The key problem of internal control dealt with by the informal courts continued to be the aintenance of big men's control over the means of production and reproduction. Of the 107 cases observed by Berndt mostly between 1951 and 1953, nearly two-thirds (69) concerned women, the largest number relating to adultery by women (29) and husband desertion (16). [149] In the informal courts, women generally continued to have no legal rights, although a husband could be punished for assault on his wife. [150] But, Pax Australiana, the operation of the informal court and the ubiquity of the new officialdom, meant that one of women's most useful defences, flight to and absorption by a hostile or at least powerful, neighbouring community, no longer afforded any protection. With colonisation and informal courts, the power of big men and male hegemony locally, formerly constrained by the men's house and limited in its range, now outgrew the village. With the demise of the men's house, and in pursuit of women, political control grew beyond the village. With this occurrence, the grounds were laid for the big men's co-option of a far more powerful instrument, the Local Government Council.

Articulation at the level of the political

With colonial intrusion, the dominant domestic groups and embryonic classes in combination with the agents of the colonial state, set in motion a process which at one and the same time contained a conserving and dissolving aspect. The balance between conservation and dissolution was determined both by the nature of the colonial capitalism and by the nature of the impacted social formations.

In the case of the former tendency, key determinants in the process were which fraction or fractions were dominant within the metropolitan ruling class; the relationship between that class or fraction and those in other metropoles; and the strength or weakness of the colonial state apparatus, and its relation to the agents of colonial capitalism. Such agents, variously missionaries, prospectors, planters and colonial officials, themselves displayed differing and changing degrees of dependence upon the colonial system.

Concerning the nature of the impacted formations, centrally significant were the contradictions within the pre-capitalist modes, and the extent to which these were resolved, mediated or sharpened in the process of articulation. The organisation, utilisation and preservation of land, labour and fertility, the means of production and reproduction, gave rise to a set of fiercely antagonistic social relations within communities in the form of male-female antagonism and between communities in the form of warfare and to the impoverishment of some men by others.

When peace was enforced between highlands villages, one set of these contradictions was resolved. The internal implications of these resolution were immense, and pacification probably exerted a strong disintegrating influence on the village. No longer was it so strictly necessary for people to live in villages. No longer could fear of enemies continue to legitimate hostility to women. No longer was recruitment and absorption of warriors of such vital importance, nor their attraction to the polity through displays of riches and the obtaining and retention of surplus land. No longer was that part of the surplus invested in alliances and bribes required. Peace, far from conserving and redeeming a blighted and doomed social system, struck at the deepest roots of an agricultural production system attaining its zenith.

The centrifugal forces within the village political economy unleashed by pacification, were held in check by colonialism too. Villagisation, the census, and roads, all served to promote conservation and centralisation. Pacification froze land holdings, and reduced the possibility of pioneering ventures by disgruntled would be big men. Land and village became synonymous, and the only possibility of obtaining the former was as a member of the latter. In fact, not only did colonialism deflect some pre-colonial disintegrative tendencies, it increased the size of the village polity, and quickened the tendency toward the centralisation of wealth within and between villages.

The agents of colonial control promoted the conservation of the village for other reasons. Australian colonialism in Papua New Guinea was not equipped to handle the massive breakdown in social order, which could have eventuated if the village had no longer functioned as a social and political entity. The missions, too, supported its continuance, since the Lutherans particularly depended upon it as a central part of their strategy of evangelisation. Reed rather bluntly stated that the missionaries wanted Papua New Guineans in their villages 'so that they could receive the preachments of their teachers'. [151]

Similarly, the colonial state quickly realised the potential of the rich and heavily populated highlands, whose fertile valleys and favourable climate appeared to be ideally suited for white settlement, and because of the 'availability of a teeming potential labour force'. [152] The census forms carried by the kiaps required 'a precise statement of the number of 'able-bodied' persons in the labour force'; [153] a labour force, the continued well-being and reproduction of which depended upon the conservation of the village.

Conquest destroyed the men's house which brought together those with effective military, economic and political power in the village and kept them apart from those whose surplus they controlled. By its very existence it embodied power, sexual separation and, importantly given emerging inequalities between men, solidarity. As well it housed the symbols of that power, and was the locus of the enactment and transmission of male rituals. The nama cult or its equivalents, centred on initiation and preparation of boys for warfare, was also the supreme expression and justification of men's solidarity and unity, and their control of fertility, land and labour.

Christianity aided this process of dissolution and dimunution of the men's house and sought to replace it in the construction and maintenance of ruling ideas, with the church. But the men's house played more than the role of mediation of the external contradictions of survival in warfare, it was also vital in the control and regulation of labour power, and the mediation of the contradictions that emerged from this. The church, capable of making grounds on the level of the 'purely ideological', was incapable of acting as a mediator in spheres of production relations within the political space of the village.

The evangelists were also unable to match the political muscle of an official. With skill, a little luck, and if necessary some bribery, a luluai could be backed by the guns of the warriors of the state. The best an evangelist could do was to threaten expulsion from the congregation, a severe act during the time of perpetual hostilities, but one of much less consequence once the blessed peacemakers had succeeded.

Nonetheless, pacification and the missions, tolerated, even sought after, for the economic advantages they brought, were forces strong enough to terminate the men's house. If anything, the male/female contradiction increased with colonialism, given the greater product required of women who could perceive a marked reduction in the labour time of their men. The big man, pivot of the men's house, had most to lose by its demise, since he exerted control over more productive resources than any other. Pacification and the informal court came to his rescue, the former making apprehension of female offenders more certain, and the latter sustaining a new and equally efficient theatre of social control. The informal court was as indigenous as its convenors and like them appeared in colonial garb. The warriors who previously performed the tasks of inquisition, judgement and retribution, were now hatted or clothed in navy blue, and commanded far more authority than they formerly did as bowmen or spearmen.

Tendencies toward the concentration of wealth and power were well evident before conquest. The price for even more power, strength, followers, and wealth and greater permanency, was the loss of some degree of independence. This many big men paid gladly. In striking the bargain, the informal court became something of a captive institution. At one and the same time, it sustained the social relations necessary for production and reproduction by maintaining and demonstrating the power of the officials over other men, and of men over all women, and it acted increasingly as a conduit for colonial control.

Given that Australian colonialism had destroyed the men's house, and that it required the continuation of the village to provide labour and maintain social control, it was not remotely possible for it to seriously challenge, let alone destroy, the institution of the big man. Many big men survived conquest with their personal political power intact, their wealth enhanced, and their control over village labour maintained. Konangil, LuBiso, Makis and other big men that accepted 'the hat' of the colonial village official would have invented local government and even cash cropping had they not been given them so readily.

Notes

[1] M.J. Leahy and M. Crain, The Land That Time Forgot, New York, Funk and Wagnalls, 1937, p. 260.

[2] J.B. Watson, 'Introduction: Anthropology in the New Guinea Highlands', New Guinea: The Central Highlands, American Anthropologist, 66, 4, 1974, pp. 6-7.

[3] S.W. Reed, The Making of Modern New Guinea, Philadelphia, American Philosophical Society, 1943, pp. 133-7, 110-111. Some areas of

Papua new Guinea resisted colonial intrusion over a long period. In one highlands Goilala valley alone between 1917 and 1938 from 75 to 100 warriors were killed in clashes with patrols. See J. Sinclair, Kiap: Australia's Patrol Officers in Papua New Guinea, Sydney, Pacific Publications, 1981, p. 153.

[4] Leahy and Crain, op. cit., p. 219.

[5] Reed, op. cit., pp. 135-7.

[6] Leahy and Crain, op. cit., pp. 143-4; 241-2.

[7] Ian Downs, 'Kiap, Planter and Politician: A Self Portrait', in J. Griffin (ed.), Papua New Guinea Portraits, Canberra, ANU, 1978, p. 228.

[8] E.R. Sorensen, 'Socio-Ecological Change among the Fore of New Guinea', Current Anthropology, 13, 3-4, 1972, p. 362.

[9] M.J. Leahy, 'The Central Highlands of New Guinea', Geographical Journal, 81, 3, 1936, pp. 241, 244, 250.

[10] Leahy and Crain, op. cit., pp. 123, 129, 134.

[11] W.A. Ross, 'The Growth of Catholicism in the Western Highlands', Journal of the Papua New Guinea Society, 2, 2, 1968, p. 62; R. Radford, 'Burning the Spears: A 'Peace Movement' in the Eastern Highlands of New Guinea, 1936-7', Journal of Pacific History, 12, 1, 1977, p. 46.

[12] R.F. Salisbury, From Stone to Steel, Melbourne and Canberra, Melbourne University Press and ANU, 1962a, p. 115; Radford, 1977, op. cit., pp. 50-51; Downs, 1978, op. cit., pp. 233-4.

[13] F. West, 'An Australian Moving Frontier in New Guinea', in N. Gunson (ed.), The Changing Pacific, Melbourne, Oxford University Press, 1978, p. 219; W.A. Ross, op. cit., p. 62.

[14] Radford, 1977, op. cit., p. 40; Ross, op. cit., p. 62.

[15] West, op. cit., p. 220.

[16] C. Ashton, 'The Leahy Family' in J. Griffin (ed.), Papua New Guinea Portraits, Canberra, ANU, pp. 178-9.

[17] Leahy and Crain, op. cit., p. 250.

[18] Ibid., pp. 258-268.

[19] Downs, op. cit., p. 235.

[20] Radford, 1977, op. cit., p. 52.

[21] Downs, 1978, op. cit., pp. 237-8. Both the romance and the power of the kiaps, is captured in the following extract from Sinclair, op. cit., p. 53, from a chapter entitled 'On Patrol':

> The ceremony of retreat, the police dismiss and darkness falls. Cookfires glow from the dark huddle of huts; there is a slow, moving murmur of voices from the villagers, sitting outside on woven mats, as they eat their boiled bananas and coconut cream and talk over the events of this big day. Before returning to the resthouse to shower, eat the rough meal prepared by Dibob and read a while by the light of the hissing pressure lamp, I stroll around the village.
> 'Goodnight, kiap' the people softly call.

[22] M. Fortes, 'Preface' in M. Fortes (ed.), Social Structure: Studies Presented to Radcliffe-Brown, Oxford, 1949.

[23] Downs, op. cit., p. 232.

[24] Fortes, op. cit., p. vi.

[25] Downs, op. cit., p. 225; Reed, op. cit., Appendix III; J. Wilkes (ed.), New Guinea and Australia, Sydney, Angus and Robertson, 1958, Table 6, 217; Table 61, 231.

[26] R. Radford, 'Missionaries, Miners and Administrators in the Eastern Highlands', Journal of the Papua New Guinea Society, 6, 2, p. 87.

[27] *Ibid*., pp. 88-9.
[28] *Ibid*., p. 90.
[29] Radford, 1977, *op. cit*., p. 43.
[30] Leahy and Crain, *op. cit*., pp. 48, 52.
[31] W.A. Ross, 'Ethnological Notes on Mt. Hagen Tribes Mandated Territory of New Guinea', *Anthropos*, 31, 1936, p. 353.
[32] Ross, 1968, *op. cit*., p. 59.
[33] *Ibid*., pp. 61, 62.
[34] Leahy and Crain, *op. cit*., pp. 162, 165.
[35] For a description of the process see Leahy and Crain, *op. cit*., pp. 76, 96, 97-8.
[36] I. Downs, *The Australian Trusteeship in Papua New Guinea 1945-1975*, Canberra, Australian Government Publishing Service, 1980, p. xvii.
[37] Reed, *op. cit*., p. 218.
[38] The Okapa patrol post, for example, was staffed for four years by a native constable. See J.F. Bouchard, 'The Impact of Roads on the Monetary Activities of Subsistence Economies in the Okapa Region of Papua New Guinea', *Department of Geography Occasional Paper Number 4*, Port Moresby, University of Papua New Guinea, 1972, p. 34.
Sinclair *op. cit*., p. 85, noted Sergeant Siria who as a first year constable had been placed in charge of 'an isolated police post in the newly opened Central Highlands where a lot of blood was shed'.
[39] Reed, *op. cit*., p. 169.
[40] Downs, 1980, *op. cit*., p. 155.
[41] Sinclair, *op. cit*., p. 45.
[42] Reed, *op. cit*., p. 213 note 5.
[43] Downs, 1978, p. 237.
[44] R.M. Berndt, *Excess and Restraint: Social Control Among a New Guinea Mountain People*, Chicago, University of Chicago, 1962, p. 334.
[45] *Ibid*., p. 359.
[46] H.I. Hogbin, 'Government Chiefs in New Guinea', in M. Fortes (ed.), *Social Structure: Studies Presented to A.R. Radcliffe-Brown*, Oxford, 1949, p. 200.
[47] Berndt, *op. cit*., p. 349, note 18.
[48] M. Alpers and D.C. Gadjusek, 'Changing Patterns of Kuru: Epidemiological Changes in the Period of Increasing Contact of the Fore People with Western Civilisation', *American Journal of Tropical Medicine and Hygiene*, 14, 5, 1965, p. 860.
[49] J. Kuder, 'The Lutheran Mission in New Guinea Today', *International Review of Missions*, 41, 1952, pp. 310-323.
[50] G.F. Vicedom, 'The Growth of the Lutheran Church in New Guinea Today', *International Review of Missions*, 45, 1956, p. 299.
[51] *Ibid*.
[52] *Ibid*., p. 927.
[53] Kuder, *op. cit*., p. 321.
[54] Vicedom, *op. cit*., p. 302. For a description of the process at work, see H.I. Hogbin, *Transformation Scene: The Changing Culture of a New Guinea Village*, London, Routledge and Kegan Paul, 1951, p. 236.
[55] Ross, 1968, *op. cit*., p. 61.
[56] *Ibid*.
[57] *Ibid*., pp. 62, 63.
[58] O.C. Fountain, 'Some Roles of Mission Stations', *Practical*

Anthropology, September-October 1971, p. 200. Hogbin, op. cit., p. 199, noted in 1951, religious 'societies which supplemented their revenue by conducting planting and trading concerns ... are largely maintained by cheap native labour'.

[59] O.C. Fountain, 'Religion and Economy in the Mission Station-Village Relationships', Practical Anthropology, 13, 2, 1966, p. 54.

[60] West, op. cit., p. 217.

[61] Radford, 1977, op. cit., p. 44.

[62] West, ibid.

[63] Reed, op. cit., p. 240.

[64] H. Aufenanger, 'The War-Magic Houses in Wahgi Valley and Adjacent Areas (New Guinea)', Anthropos, 54, 1959.

[65] Berndt, op. cit., p. 382. The indigenous theological adaptation of the bible story provides a fascinating revelation of the political realities of village life under colonial rule. Pontius Pilate was portrayed as a district officer; Roman soldiers as police; the Jewish authorities as luluais; God and Satan as competing big men with God appearing in coats, trousers and boots; and the Garden of Eden referred to as a pre-colonial time when white rule was unnecessary because the people had not 'fallen'. See Hogbin, op. cit., pp. 239, 241 note 1, 247.

[66] M. Reay, 'Generating Political Conflict: Some Consequences of Economic Exploitation of the New Guinea Highlands', Anthropol-oligical Forum, 3, 1973-74, p. 298.

[67] Reed, op. cit., p. 238.

[68] K.E. Read, 'Missionary Activities and Social Change in the Central Highlands of Papua New Guinea-1', South Pacific, 5, 11, p. 234, 1952.

[69] L.L. Langness, 'Ritual, Power and Male Dominance in the New Guinea Highlands', Ethos 2, 1974, p. 210. S. Glasse, 'The Social Effects of Kuru', Papua New Guinea Medical Journal, 7, 1964, p. 38; R. Glasse, 'Cannibalism in the Kuru Region of New Guinea', New York Academy of Sciences Transactions, Series 2, 29, p. 749.

[70] Sinclair, op. cit., p. 199.

[71] Aufenanger, op. cit., p. 36.

[72] S. Glasse, op. cit., p. 36.

[73] Read, op. cit., p. 234.

[74] Reed, op. cit., p. 111.

[75] Vicedom, op. cit., p. 304.

[76] Kuder, op. cit., pp. 320, 322.

[77] Hogbin, 1949, op. cit., p. 197.

[78] Alpers and Gadjusek, op. cit., p. 855.

[79] Leahy, op. cit., p. 233, 251.

[80] Salisbury, 1962a, op. cit., pp. 118, 121.

[81] R.F. Salisbury, 'Early Stages of Economic Development in New Guinea', Journal of the Polynesian Society, 71, 1962b, p. 329.

[82] Salisbury, 1962a, op. cit., p. 119. R.F. Salisbury, 'Despotism and Australian Administration in the New Guinea Highlands', New Guinea: The Central Highlands American Anthropologist, 66, 4, 1964, p. 7; Sorenson, op. cit., pp. 368-9.

[83] Salisbury, 1962b, op. cit., p. 329.

[84] Salisbury, 1962a, op. cit., pp. 116, 122.

[85] Ibid., pp. 115, 124.

[86] Ross, 1936, op. cit., p. 353.

[87] P.L. Newman, Knowing the Gururumba, New York, Holt, Rhinehart and Winston, 1965, p. 12.

[88] K.E. Read, 'Land in the Central Highlands', South Pacific, 6, 7, 1951, p. 444, ff.

[89] Sorenson, op. cit., p. 362. It is doubtful whether villagers made great distinctions between missionaries and anthropologists. After he had left Busama, the anthropologist Hogbin was beseeched to 'come home and be our missionary. Never mind about the Bible'. Hogbin, 1951, op. cit., p. 235.

[90] Leahy and Crain, op. cit., p. 168.

[91] Ibid., p. 221.

[92] Salisbury, 1962b, op. cit., p. 328.

[93] Downs, 1978, op. cit., p. 238.

[94] M. Reay, 'Present Day Politics in the New Guinea Highlands', New Guinea: The Central Highlands, American Anthropologist, 66, 4, 1964, p. 248.

[95] Salisbury, 1962b, op. cit., p. 328.

[96] Sorenson, op. cit., pp. 364-5; Alpers and Gadjusek, op. cit., p. 856.

[97] S. Glasse, op. cit., p. 47.

[98] Sorenson, op. cit., p. 362; E. Sorenson and D. Gadjusek, 'Nutrition in the Kuru Region I. Gardening, Food Handling and Diet of the Fore People', Acta Tropica, 26, p. 285.

[99] Newman, op. cit., p. 19.

[100] Salisbury, 1964, op. cit., p. 3.

[101] D.C. Gadjusek and L.H. Reid, 'Studies on Kuru IV. The Kuru Pattern in Moke, A Representative Fore Village', American Journal of Tropical Medicine and Hygiene, pp. 630, 636.

[102] Salisbury, 1962a, op. cit., p. 116.

[103] Salisbury, 1964, op. cit., p. 6, note 16.

[104] Ibid., pp. 6, 7.

[105] Salisbury, 1962a, op. cit., p. 138.

[106] West, op. cit., p. 244.

[107] K.E. Read, 'Leadership and Consensus in a New Guinea Society', American Anthropologist, 61, 1959, p. 436.

[108] Reed, op. cit., p. 218, note 5.

[109] Alpers and Gadjusek, op. cit., p. 865.

[110] West, op. cit., pp. 224, 225.

[111] Reay, 1964, op. cit., p. 247.

[112] Hogbin, 1949, op. cit., p. 193.

[113] Ibid., 245.

[114] Downs, 1981, op. cit., p. 1.

[115] A.M. Healy, 'Agriculture in Kenya and New Guinea', Australian Quarterly, 23, 2, p. 169.

[116] Salisbury, 1962a, op. cit., p. 136.

[117] L.L. Langness, 'Notes on the Bena Council, Eastern Highlands', Oceania, 33, 3, 1963, p. 154.

[118] Berndt, op. cit., p. 319; Hogbin, 1946, op. cit., p. 44.

[119] J.A. Barnes, 'Indigenous Politics and Colonial Administration with Special Reference to Australia', Comparative Studies in Society and History, 2, 1959-60, p. 145.

[120] Reed, op. cit., pp. 165, 240, 218.

[121] J.R. McAlpine, 'Eastern Highlands District', in P. Ryan (ed.), Encyclopaedia of Papua New Guinea, Melbourne University Press, 1972, p. 288.

[122] West, op. cit., p. 224.

[123] Radford, 1977, op. cit., p. 47.

[124] Ibid, p. 52.

[125] Berndt, op. cit., p. 382.

[126] Sinclair, op. cit., p. 53.
[127] Reed, op. cit., p. 232, 260, 264. Or caused the most fear? Hogbin, 1951, op. cit., p. 286 noted 'innumerable cases ... of village officials being struck and village women raped' by native police.
[128] Berndt, op. cit., p. 321.
[129] C. Simpson, Adam in Plumes, Sydney, Angus and Robertson, 1954, p. 16, note.
[130] Bouchard, op. cit., p. 40.
[131] R.G. Ward et al, Growth Centres and Area Improvements in the Eastern Highland District: A Report to the Central Planning Office, Canberra, Department of Human Geography, Research School of Pacific Studies, ANU, 1974, p. 19.
[132] Downs, 1978, op. cit., p. 246.
[133] R.J. Southern, 'Road Transport in the New Guinea Highlands', Department of Geography Occasional Paper, Number 6, Port Moresby, University of Papua New Guinea, 1973, p. 23.
[134] Anon., 'Progress in the Lufa Area', Australian Territories, 3, 1, 1963, p. 43.
[135] Ibid.
[136] Sorenson, op. cit., p. 366.
[137] Anon., op. cit., p. 43.
[138] Berndt, op. cit., p. 383, note 4.
[139] Downs, 1980, op. cit., p. 149.
[140] Berndt, op. cit., p. 313.
[141] Ibid., p. 318.
[142] Downs, 1980, op. cit., p. 150.
[143] Ibid., p. 31.
[144] Calculated from Berndt, op. cit., pp. 331-375. Further on sexual antagonism and the men's house, see M. Donaldson, 'Contradiction, Mediation and Hegemony in Pre-capitalist New Guinea: Warfare, Production and Sexual Antagonism in the Eastern Highlands', pp. 435-457 in Melanesia: Beyond Diversity, vol.2, R. May and H. Nelson (eds), Research School of Pacific Studies, Australian National University, Canberra, 1982.
[145] Reed, op. cit., p. 177.
[146] Berndt, op. cit., p. 375.
[147] Berndt, ibid.
[148] Ibid., pp. 323, 341, 353.
[149] Ibid., pp. 328-330.
[150] Ibid., pp. 325, 364.
[151] Reed, op. cit., p. 106.
[152] Reay, 1973-4, op. cit., p. 297.
[153] Ibid.

4 Colonialism and the development of capitalism

The spread of wage labour

Gitlow, relying mainly on the testimony of the missionary Ross, provided evidence in accordance with that of Vicedom and Tischner, [1] that there existed in the Mt. Hagen area a number of what he called 'serfs'. These men were essentially male women, for like women they enjoyed no property rights and needed to enter into a relationship of dependency to secure the means of livelihood. [2] Gitlow suggested that they may have been refugees or the sons of fathers unable to provide bridewealth, and as such provided day to day labour in exchange for protection and security. [3] Ross estimated that 10 percent of the population of the Hagen area belonged to this class. Ninji the most influential of the Hagen big men, appointed bosboi by the kiap, had five 'serfs' and perhaps 10 wives. [4]

The exchange of labour in return for goods or services, although restricted, was not new to the people of the central cordillera. The missions in particular (as noted in chapter 3) were quick to capitalise on this. Prior to the war, Ross could purchase half a day's labour for half a sheet of newspaper. [5] At Kainantu, 10 years before the introduction of cash, the mission stations at Kambaira, Omarua and Raiapinka and the miners' camps at Yonki, Omarua and Barola were substantial and regular users of indigenous labour. [6] As early as 1939, the villagers around the Kainantu patrol post were refusing work, and the continuing high rates of absenteeism meant that labour had to be brought in from beyond the area. [7]

Many would have gained their first taste of labour discipline carrying between 16 and 23 kilos for twist tobacco and keep, to the value of perhaps 2 pounds per week. [8] Idriess suggested that each miner required at least 10 carriers, [9] but perhaps a greater number of villagers would have experienced state rather than private enterprise patrols. Carriers on such ventures were generally changed at each village, and even if the villagers were unwilling, 'there was no escape for them; the work of the government had to proceed'. [10] On the patrol through Menyama in 1950, Sinclair lead 157 carriers with 2-1/2 tonnes of gear. [11]

Carrying for patrols was difficult to avoid, but it was harder to escape from road building duties. The only way to dispense with human porterage in the long run, was through its intensification in the short term. The road system through the Eastern Highlands, which ran from Gusap to Goroka by 1953, was almost solely the product of unaided human labour. Gangs of up to 10,000 men drove the road up and over the Daulo Pass to link the Asaro Valley with Simbu, without the benefit of machinery or even protective clothing in high altitude conditions which were not only wet and cold but hard and dangerous. [12] Less grandly, but just as impressively, 'hundreds of tribesmen' hauled 600 tonnes of hardwood for up to 14 miles to suspend a bridge over the Mabimoka River, one of hundreds built similarly by highlanders during the construction of the colonial highway. [13]

The system of rostering all the able-bodied men and sometimes women to undertake road work one day per week, was sufficient for road maintenance, but not for road construction. With the reluctance of villagers to undertake the heavy and boring work, roadworkers were compelled to live in rough barracks near the road, with the result that they were absent from villages and gardens for long periods. [14]

For those who sought to avoid compulsory carrying or road work, or who were apprehended at other more serious crimes, labour discipline was instilled in gaol. Prisoners built District Officer Sinclair a new office and store, the framework for which was 'cut in the mountain forest far from the station and hauled in bodily by gangs of sweating prisoners'; [15] the difference between them and their free roadworking compatriots perhaps being that in <u>kalabus</u> the food supply was more regular and more plentiful.

The major users of native labour in the highlands remained the missions and government until the middle 1950s. A mission station, a government post and plantation, exerted a similar effect on the area in which they were located. [16] The missions, by the 1930s, were working 4,000 uindentured labourers in New Guinea, mainly in the cultivation of 8,600 hectares of copra plantations, [17] and each patrol post, 12 to 700 workers. [18] The largest single concentration of wage labour was the Wau goldfields - 7,000 Papua New Guineans were working there in 1938. [19] They were joined, before the war, by a small number of highlanders who were recruited illegally, most of whom 'quickly died'. [20]

Highlanders' lack of immunity to white disease had dreadful consequences too within the region. An influenza epidemic swept the eastern part of the region in 1936, [21] and a dysentry epidemic hit Hagen in 1938, the year the first <u>kiap</u> took up residence. [22] Hagen was next afflicted by eczema, dysentry and meningitis, followed by bacillary dysentry introduced there by an American serviceman in 1943. [23] An estimated 5,000 central highlanders died as a result, [24] up to 20 per cent of the adult population in some villages. [25]

The 1943-44 dysentry epidemic, probably more than any other factor, prevented the uncontrolled exploitation of highlands' labour. The reparations paid elsewhere to villages which had sustained damage through the war were paid in the highlands to the Department of Public Health instead, supposedly because the highlanders were not acquainted with money. [26] Having spent about a quarter of a million pounds on the control of disease, the government was concerned that returned workers might cause similar outbreaks. In 1946, a Labour Ordinance was enacted forbidding the employment of highlanders at, or their passage through, altitudes less than 3,500 feet. [27]

By the end of 1944, with some territory still under Japanese control, 21,000 New Guineans were under contract, [28] of whom thousands had not been home for three years, and many had been working for up to six. [29] In October 1945, all existing indentures were cancelled 'on the legal ground that they had been made under compulsion and on the humanitarian ground of the strain placed on the indigenous inhabitants by wartime demands'; [30] strain that was apparently leading to 'growing unrest'. [31] By February 1946 the labour force had fallen by 4,100 at a time when the post-war reconstruction of agriculture and industry required a greater labour force than ever. [32]

By 1948, the work force had risen to nearly 26,000 in New Guinea, but there was estimated to be a shortage of 24,000 workers. [33] In 1949 a deputation from the Papua Planters' Association, informed the colonial government that the workforce was 57 per cent below strength, a situation brought about in part by the large number of workers required for minerals exploration and extraction. [34] 'Employers' eyes began to turn to the one great untapped source of labour in the territory ... the central highland valleys ...'. [35]

Gunther, the Director of Public Health, took 'personal responsibility' for the government's decision in 1949 to permit highland's labour to offer for employment in coastal areas. [36] In 1949, New Guinea Goldfields Ltd. was given permission to employ 210 workers from Bena for its mines at Wau. [37] The experiment was apparently successful, for in the same year permission was granted for the employment of highland workers in Port Moresby, Lae, Rabaul, and Madang. The medical provisions were strict, and involved a medical examination prior to, and a period of quarantine following migration, at Goroka. [38] The Highlands Labour Scheme went into operation in January 1950, and by the end of that year, over 3,000 workers had been flown from Goroka to the coast. [39]

The existence of a wartime aerodrome in Goroka, and the centralisation of labour recruitment there, were interrelated. As the Highlands Labour Scheme developed, air services dramatically increased such that by the early 1950's, almost a plane a day was leaving Goroka. [40] A number of them flew for Jim Leahy's Mandated Airlines Agency and were fuelled by his Vacuum Oil Agency. [41] The provision of labour for the coastal plantations provided a means of capital accumulation for the small but growing band of white entrepreneurs attracted to Goroka by the easy availability of labour, the pleasant temperate climate, and air access to and from the coastal towns.

By August 1951, more than 10,000 highlanders were indentured under the scheme, 90 percent of them from the Simbu Valley. [42] In the Upper Simbu census division in 1953-1954, 21 out of 54 'clans' were over-recruited, that is more than one in three of the adult men were absent. [43] Simbu's relegation to the position of a labour reservoir was underway.

Labour recruitment posed the village leadership with severe problems, not only in the reproduction of the means of production, but also with the reproduction and maintenance of social relations. The production of a reasonably regular surplus was a precondition of the continued existence of the agricultural mode of production dominant in the Eastern Highlands (see chapter 2). Surplus product was necessary not only because the climate made settled agriculture something of a risk in higher parts of the region, but also because the most basic tool in the Eastern Highlander's kit bag, the axe or adze, generally had to be obtained by trade. Similarly, agricultural labour, in the form of women, could only be obtained through the payment of bridewealth, dependent on the production of a tradeable surplus.

The immediate economic problems posed by indentured labourers were resolvable by the wages they returned to the village. The economic surplus was now (partly) created outside the village, and could be used to purchase the basic means of labour, the axe, from the tradestore. This in turn alleviated male labour shortage by substantially reducing the labour time necessary to complete male tasks. But this solution to the problem of labour absence, also created dependence on money, and initially therefore on wage labour. As Hogbin commented in 1951,

> The peoples of New Guinea ceased to be independent, however, from the moment when they first began to appreciate the worth of steel ... if the natives of today were deprived of money to buy axes they would starve. [44]

This new dependence on money created severe problems for the holders of political power. The men who sold their labour power and whose wages came to constitute a vital element in social reproduction controlled the most crucial means of labour in the village political economy in a manner that had not been possible prior to the intrusion of capitalism. Formerly the big men had dominated the political and economic life of the village, through their control of land and labour, the means of production and reproduction. Other goods were now controlled by younger men whose absence in the obtaining of these goods in fact created the need for them: the more young men were absent, the more the village depended upon their wages leading to a deepening circularity and the creation of permanent labour reservoirs.

The reintegration of returning workers was a problem. The big men, though, had major advantages for they retained their control over the allocation of village land and largely over women's agricultural labour; they retained and even strengthened their ceremonial and trading linkages with other big men; bridewealth inflation tended to work in their favour; and they retained control over the values and apparatus surrounding the ideology of male supremacy.

The elders in post-war Busama bemoaned the demise of male initiation and the learning period associated with it, for this had been the main means by which respect for authority was instilled. Bumptiousness and insolence characterised the behaviour of recently returned labourers; 'The villagers called them 'big heads' ... This exactly described them'. [45] In Siane, the first indentured labourers left in 1950 and returned in the middle of 1951. A second larger group had returned by January 1953, by which time a 'returning ritual' had become traditional, the effect of which was to assist the reabsorption of the workers into the village, and to relieve them of their cash and goods. [46] Such integrative techniques seemed to prove successful, for in 1958, West commented that 'so far there has not been a serious problem of detribalisation and the frustrations of returned labourers have not caused serious trouble'. [47]

Such mechanisms were backed by the sanctions of the colonial state. When the government was 'confronted by sullen or uncooperative repatriated labourers, its short term solution [was] to support ... the _luluais_ with the penal sanctions of the Native Administration Regulations and to trust that these headmen [could] deal with the situation by use of their authority and influence'. [48]

The _kiaps_ too were forced to rely even more on the appointed officials because of the increased workload thrust upon them by the Highlands' Labour Scheme. Downs noted 'our most unpleasant duty was the apprehension of deserters from distant labour contracts. There was no such thing as a reasonable excuse'. [49] On the one hand the kiap would support the luluais against the 'presence of these wealthy labourers [which] obviously upset[s] the traditional balance', [50] and on the other hand would rely upon the officials to report and perhaps even apprehend deserters.

Yet the feature which most importantly accounted for the reabsorption of the returnees was the appropriation of their goods through the 'reciprocal' exchange relations existing within village political economies prior to and since white intrusion. Wealth still flowed into the possession of those most active in exchange activities, such that 'one could say that the big men's authority was effective in enforcing the distribution of valuables'. [51]

A young man could normally anticipate making payments amounting to 28 gold lip shells and over 30 other valuables in the twenty years between his marriage and his death. In addition, he had to kill more than 30 pigs. Even a returned labourer could not possibly finance payments on this scale. To ensure the obtaining of full social adulthood, young men had to indebt older and richer men to them, so that they could rely upon their 'generosity' at the appropriate times. This long term dependence on big men was an important factor in their retention of control over young men, and it forced returned workers to distribute the new valuables which in theory might have threatened the big men's control of land, labour and brides. Most notably, it was possible 'for an unruly youth to be brought to his senses by a reminder that he may find himself condemned to a permanent state of bachelordom'. [52]

Salisbury suggested that about half the goods brought home by the returnees found their way into the hands of the luluais, big men and bosbois. The richest man in Siane was a stay at home luluai. [53] This tendency was further accelerated, according to patrol officers' reports, by bridewealth inflation. [54]

Reabsorption of workers returning from indentured labour on coastal plantations was reasonably assured by 'new' ideological mechanisms and by the continuation of economic inequalities and their supports. In addition, such was the concern of the colonial state with the possibility of supposed detribalisation, [55] that workers by law had to be returned to their villages. Both the big men at home who frequently worked for small material rewards to assist recruitment, and the workers away, knew that when the period of indenture was over, the indentured workers would be returned.

But even with casual workers in the cities, village linkages, if not control, were maintained. Workers in the towns would tend to live with and associate with workers from the same or allied villages and would provide support and assistance for newly arrived migrants. [56] Among the Siane in Port Moresby in the early 1970s, 80 percent of the members of a migrant group comprised fellow villagers. [57] Indeed, most workers migrated with the idea of returning to their villages. Public servants in particular had a considerable stake in returning to become, in their turn, 'traditional' and wealthy village leaders, and to achieve this, their linkages with the village had to remain intimate. [58]

West noticed, as early as September 1952, 'a marked decrease in enthusiasm to engage for coastal labour' because when the worker returned home 'the prestige he hoped to have gained ... fail[ed] to materialise'. [59] This lack of enthusiasm increased as coffee was introduced and began its rapid but geographically and socially uneven spread through the highlands. Although cash cropping was seriously underway in the Eastern Highlands by 1955, it was not until the late 1950s that the first major harvests were being brought in. By the late 1960s, plantation workers were receiving 25 cents per day plus board and lodging, but some estimates suggested that the return for an average coffee grower was $1.80 per day worked, or $7 a week. [60]

The growth of coffee farming in the Eastern Highlands forced the recruiting centres to move westwards from Goroka to Mt. Hagen and Wabag, and then to Mendi. [61] As early as 1957, fringe areas were beginning to function as ancillary labour reservoirs, particularly (southern) Simbu where land shortage and population density had accelerated the development of sweet potato agriculture, and the southern areas of the Eastern Highlands where pre-ipomean forms of agriculture were still being practised when colonialism intruded. Thus the areas to benefit from the introduction of cash cropping were those which had left taro production far behind, but did not yet possess such intensive sweet potato cultivation as Simbu. Hagen, Kainantu, and particularly Goroka, became centres of coffee plantation and peasant commercial agricultural development. Both forms of cash cropping were rooted in pre-existing pre-capitalist inequalities and drew workers from the 'fringes' to the central belt. [62]

The introduction of local government

Following the resumption of civilian administration at the War's end, the administrator, J.K. Murray, concerned at the supposed failure of the luluai system, convened a series of conferences to advise the colonial government on what changes should be made to Territory ordinances to legitimate the informal courts and village councils that were increasingly a feature of indigenous political life. As Hogbin commented, 'creating a formally recognised council-cum-court ... will not be in any sense a revolutionary action. Little change will be necessary in the conduct of the [existing] gatherings ...'. [63] Hogbin recognised that the men's house was the main juridical and ideological instrument in the village polity prior to the advent of colonialism, and that it might therefore 'be regarded as a rudimentary form of court'. [64] Given the widespread existence of 'informal courts' which combined elements of the precapitalist political system and those imposed by the colonial state, Hogbin advised that 'with such excellent foundations already laid, the obvious course is to give the native institutions official recognition ...'. [65] This recognition was forthcoming, and in December 1949 legislation for the creation of village councils was passed.

The personnel of the 'new' institutions was remarkably congruent with that of the 'old' for 'in most cases the councillor was a luluai'. [66] The paramount luluai Tahu at Kaiapit became the council president, and 'completely dominated' the operation of the Bukam council, [67] and Paramount luluai Nason Tokiala became council president on the Gazelle. [68]

Even in the 1970s, the continuity remained evident. Of 40 councillors in the Dei council area in the Western Highlands, 27 had been bosbois, luluais or tultuls. Equally significantly, in 1968 20 out of 35 were big men, and in the early 1970s, 27 out of 40 were big men or sons of big men. [69] Even this may underestimate the continuing power of the big men, for some councillors, while not big men or their sons, were certainly their proteges and owed the council position to the support of a big man. [70] The president of the Dei council was the son of a big man, and by 1964 had acquired a tradestore, followed by a truck in 1965 and thence coffee buying and the establishment of a small tea plantation. [71]

Just as the offices of big man, luluai and councillor tended to overlap and reinforce one another, so did the behaviour of the individuals concerned. As Strathern commented, 'An influential councillor can gain benefits for his group, just as in the indigenous local system a big man can arrange alliances that benefit himself and followers'. [72] As a councillor explained, 'councillors are a new type of kiap', [73] a view shared by the electors, who regarded a councillor as a black-skinned man who bossed them, as a spokesman for the colonial government, as a tax collector, as a magistrate and legislator, and as an 'unusually affluent' member of the community. [74]

Certainly, the tasks of the councils differed little from those of the 'informal courts'. They were empowered under the 1949 legislation to enforce customary obligations and requirements except in so far as custom might conflict with colonial law or be 'repugnant to the principles of humanity' [75] and they took over most of the powers formerly held by luluais. [76] If anything, the councils increased the power of the big men who were required 'to carry out various functions additional to those they had traditionally exercised in their own society'. [77] Of course the councils continued to be concerned with the control of women, and since those who paid tax to the council were eligible to vote in council elections, in the Minj council area, only men were required to pay tax. [78]

The colonial state regarded the councils as systems of administration, administrative agencies, and as means of controlling the indigenes as surely as if they had been under the luluais. [79] The Minister for Territories, Paul Hasluck, remarked in 1958,

> The standing instruction is that all officers are to watch for any awakening sense of political interest and when that interest awakens to be ready to find an outlet for its expression. In keeping with this approach, from time to time local government councils are formed in villages or groups of villages. [80]

The fusion of colonial and local political control was strengthened by the appointment of an 'advisor' (usually a kiap) to the council. In 1970, only 3 of the 145 councils were without one. [81]

Hasluck's concern to develop an economy that could 'sustain the burdens of its own administration and provision of services', [82] was translated in practice to the requirement that councils be financially self-supporting. This necessitated the grouping of a number of villages to form one council area, a further acceleration of the centralising tendencies noted earlier. This led, as Hasluck noted in 1962, to a shift in 'immediate loyalties ... from the clan and village group to the local council' [83] such that local government councils 'replaced villages as authorities'. [84]

As formerly, the increasing centralisation of political power was accompanied by its concentration. Under the luluai system, New Guinea retained one colonial indigenous official per 100 villagers (see chapter 3). Under the council system, the number of councillors per person decreased, as the system became increasingly well established and councils were encouraged to amalgamate in the interests of efficiency and economy.

Table 4.1
Local Government Councils, Papua New Guinea, 1955-1963

	Number of Councils	Number of Councillors	Persons per Council	Persons per Councillor
1955	6	141	5,000	212
1958	15	379	6,066	240
1960	23	657	7,304	255
1963	50	1,518	10,140	332

Source: Calculated from Downs 1980, p. 136.

Expatriate political involvement too, grew apace with the development of indigenous political forms. The Highland Farmers and Settlers Association had several branches through the highlands and was seen by villagers as being the whites equivalent of the councils. [85] Obsessed with the notion of 'partnership', which would ultimately result, according to assistant administrator Gunther, in 'successful farmers from all the various communities working congenially side by side', [86] the colonial state moved quite rapidly to introduce multiracial councils. Local government was initiated in the Eastern Highlands in 1958, [87] and in 1965, Goroka elected the first council in the country to include an urban area, which became multi-racial in 1969. [88]

The state, at the local level, however, had some bridges of its own to mend with the growing number of white planters and traders. Where relations with the planters and the kiaps were congenial, as was the case in the highlands generally, the new arrangement simply involved a formalisation of the existing status quo. That is, where whites in the highlands would anyway meet and consult as a matter of daily living, the colonial state introduced a mechanism to further strengthen and co-ordinate their interests through the creation in the late 1940s of the District Advisory Councils (DACs). [89] The DACs comprised white representatives selected by the district officer to advise him on the running of the district. By 1956, following the obvious early success of the council system, the administration decided that councillors would be eligible for appointment to the DACs also, and by the early 1960s, councillors made up the majority of their membership. [90] As the kiap came to advise the councils, so too did the councillors now come to advise the kiap. In 1968, the United Nations recommended that the councils should be given the opportunity to elect their representatives on to the Advisory Councils, and in 1971, Area Authorities were created, two-thirds of whose members were elected by the local government councils in each district. [91] These in turn became the basis of Provincial Government, a most highly centralised and powerful semi-indigenous local political institution.

The increasing concentration and centralisation of power which characterised post-war political change in the highlands was a product of and a catalyst for the emergence of classes of rich peasants and rural capitalists. By the late 1970s, these classes were firmly entrenched in the political economy of the area, and controlled the most powerful local political institution, the provincial government, the roots of which stretched back more than 30 years to the highly effective pre-capitalist political institutions of the big man and the men's house.

Kiaps, planters and the beginnings of capitalist agriculture

Reed has suggested that had the war not changed the fate of colonialism in the area, then the highlands populations would have been incorporated earlier into the colonial economy. [92] As it was, the 'opening up' of the highlands was a slow affair, receiving its main external stimuli from the Second World War, the end of the 'long trough' and the beginning of the 'second boom' in Australia, and the change of government from the conservative coalition to the Labour party in Canberra. Its main internal impetus came from the demise of the Morobe goldfields, the subsequent drop in the value of exports, and the 'over-cropping' of labour which became problematic after 1945.

Jim Taylor established the Upper Ramu patrol post (later called Kainantu) in 1932 and followed this by opening the Bena Post beside the Leahy brothers' airstrip at modern Goroka. Later in 1933 he and Mick Leahy led a joint private enterprise and government patrol from Bena to Mt. Hagen. By 1938 correspondents to the Pacific Islands Monthly were supporting Director of Agriculture, George Murray's vision of a 'second Kenya' in the highlands. One report indicated that several goldminers were preparing to plant coffee and soya beans on a commercial scale. [93]

This interest in commercial agriculture had been stimulated by the establishment of an agricultural research station in the Eastern Highlands, whose first director, Bill Brechin, arrived in the district in May 1936. His task was to establish a cinchona nursery (the commercial derivative of which was quinine) with seed that Murray had smuggled in from Java, and the proposed station was justified economically in that it would reduce imports of the only known preventive of malaria. [94] In 1937 the sale of 800 hectares in a small valley ten kilometres south of Kainantu was negotiated with the Aiyura, Onamunapa and Akenapa people. This is the site of what is now the Aiyura Highlands Agricultural Experiment Station. [95]

According to J.W. Barrie, Department of Agriculture, Stock and Fisheries (DASF) extension officer in Goroka in the 1950s, it was 'fairly evident' that coffee seed was first introduced into the highlands by Lutheran missionaries 'some years before the war', and 'the first native coffee plantings' in the highlands originated from seed or seedlings distributed by the Lutheran mission. Agricultural extension officer, R.W. Cottle, reported in July 1953, that there were 240-400 robusta trees in villages in the district.

Arabica coffee, a type far better suited to the highlands climate, was introduced by the Aiyura station from a plantation at Wau in Morobe, established by the New Guinea Department of Agriculture in 1928. By 1940 Brechin was able to export more than 60 tonnes valued at 2,060 pounds and in the following year he built a coffee fermentary to process the coffee berries to parchment stage. [96] According to Radford, 'Local villagers were impressed by the coffee and in May 1940, asked Brechin for seedlings for their own gardens'. [97] Under the wartime Australian military administration (ANGAU), coffee was opened to more Eastern Highlanders when Warrant Officer J. Armitstead in conjunction with Captain A.J. Schindler, officer-in-charge at Aiyura, introduced the Kami and Arufa people of Goroka to coffee cultivation, and Armitstead returned again in October 1945 to check on their progress. [98] Cottle (see above) said that there were about 1.3 hectares of arabica in the district in 1953, most or all of which was planted under ANGAU direction.

In contrast with the situation in the Ghanaian cocoa industry in the development of which settlers and foreign-owned plantations 'played no role', [99] foreign planters have had significant influence on the growth of Papua New Guinean coffee production. The lesson of the first crop worth 2,000 pounds was not lost on Jim Leahy who was attached to the Aiyura station for the latter part of the war. Village production preceded him, but Leahy was the first private individual to introduce the crop on a commercial scale. When in 1948, the administration allowed whites to establish plantations and other enterprises in the highlands, Leahy applied for a licence to open a trade store and establish a small coffee plantation at Goroka. [100] He noted, 'The people were still using shell. They had money from the war and Jim Taylor was trying to teach them what a pound note was worth...' [101] Indeed, Major Taylor circulated a letter from the Bena patrol post in May 1945 entitled 'Mercantile Stores and the Introduction of Coinage', [102] and in 1948, as district officer, he commenced paying Eastern Highlanders employed by the government in cash, and the Goroka native hospital began purchasing vegetables for cash rather than shell. [103]

In that same year, four kiaps, Taylor, Searson, Black and Blood took part in a rush to a new goldfield in Porgera, west of Wabag. [104] Two remained in service, but Taylor and Searson resigned to become planters. [105] Kiaps continued to leave district government and go into business. In 1952, George Greathead, the district officer who succeeded Taylor, resigned and established a passionfruit pulping industry for the Australian company Cottees, [106] and Dick Kelaart a fellow kiap from the old Morobe District, set up business in Simbu in the same year. [107] Within hours of Greathead's resignation, Downs had applied for the position only to follow Taylor into planting in 1956. [108] The highlands were opened up for applications for land lease in May 1952, and in July, Leahy sold his first coffee in Australia for about 7 shillings per pound. New planters rushed in, and the hectarage leased to whites in the Central Highlands rose from 1,947 hectares prior to 1952 to nearly three times this amount in 1954. A local land boom was underway.

In 1954 the government became the prime initiator of purchase agreements. The land was to be selected by the Lands Department, was to be surplus to the present or future needs of the indigenes, and was to be open to competitive bidding. [109] The search for 'waste and vacant' land proved, as Read [110] predicted, somewhat illusory. This, plus the weight of previous land legislation reiterated by Hasluck, effectively halted the 'land-boom', leaving in 1955, 55 white holdings in the Eastern Highlands (76 in the highlands as a whole) covering 5,928 hectares, with 820 hectares under cash crop cultivation. Forty five non-indigenous and more than two thousand indigenous workers were employed on these plantations. [111] By 1954, although the highlands plantations had only been established in the previous four years, the region already produced 251 hundredweights of coffee out of a total for the Territory of 1,981, establishing itself as the second largest coffee bearing area in New Guinea. [112] The government responded to the obvious importance of coffee in the highlands, and in 1954 the agricultural research station at Aiyura commenced a vigorous programme of research into coffee growing and coffee processing.

Rising coffee prices during the 1950s pushed up the value of the plantations, and in 1958 two plantations in Goroka were sold for 36,000 and 17,000 pounds respectively. By 1961, one third of the leases had changed hands. [113] Diversification proved profitable to the new planters, and in the last six months of 1956, D. McBeath, a planter in

Kainantu, purchased one tonne of coffee from indigenous growers in the area, more than the only other buyer in the sub-district, the Aiyura research station. At the same time a Gorokan planter, owner of the 40 hectare 'Hobe' plantation and later principal director and shareholder of the coffee buying company, Highlands Commodity Exchange, purchased the major portion of the 45 tonne peanut output of the Goroka sub-district. M. Matus of 'Memeneluka No. 1' plantation established a potato chip industry and purchased 22 tonnes of 'English' potatoes during the period. [114]

By the 1960s, considerable concentration was already evident, with Leahy and his family firm of Collins and Leahy, for example, buying out both Taylor and Searson. [115] In 1962, Leahy with his nephews, established a coffee processing, buying and marketing business in Simbu (Hi-Buy), and in 1965 he acquired 'Jascar', a property outside Kainantu comprising a 60 hectare cattle and coffee complex, plus butchers' shop, a petrol agency and a couple of trade stores in Kainantu and established the first coffee factory there.

> No one, least of all Jim (Leahy) would claim that Hi-Buy was established primarily for the welfare of the Chimbu people. It was a highly profitable business venture with the additional value, which he recognised, of assisting the Chimbu people to share in the coffee industry, thereby reducing the political vulnerability of expatriate business interests. [116]

The enterprising planters not only acted effectively economically, but they were also well organised politically. Their main voice was the Highlands Farmers and Settlers Association (HFSA), founded in the early 1950s by the first three district officers to be based in the Eastern Highlands, Taylor, Greathead and Downs, and by the major businessman in the highlands, Jim Leahy. [117] To this powerful quartet, came A.J. Schindler in 1963 who, after more than ten years as an agronomist at Aiyura, retired to grow coffee at 'Karanka' plantation between Arau and Karanka villages. [118]

The planters were concerned with three main things: the availability of adequate expertise; the securing of a stable market; and the provision of adequate infrastructure. HFSA set about tackling the first by importing Professor Y. Baron Goto, an expert in tropical agriculture and coffee production,in 1956. Goto toured all through the highlands assessing the potentiality of the coffee plantations. The assistant agricultural officer for Kainantu sub-district noted in his monthly report for February 1956, that 'following Prof. Goto's visit to this sub-district there is a general feeling of disillusionment among the settlers, as the majority of them are sitting on non-productive land'. [119] HFSA worked in close conjunction with DASF advisory services and used the central plantations of its members to hold field days, provided DASF with space on plantations on which to grow experimental plots, and encouraged DASF officers to contribute appropriate technical articles to its regular bulletin. [120]

The second problem, that of markets, HFSA set about rectifying in an energetic and straightforward way. Mainly because of HFSA negotiating and lobbying, the Australian Tariff Board introduced a requirement that Australian coffee buyers buy at least twenty five percent of their coffee imports from New Guinea in order to qualify for tariff concessions on all the coffee they imported. [121] This assured the highlands planters of a market for their product. The special protection for New Guinea coffee proved important. In 1969, N. Bluett of the Coffee Marketing Board estimated that Australian buyers imported an extra 2,500-3,000 tonnes of New Guinea coffee because of the tariff agreement. [122]

HFSA was also active in organising marketing facilities but this problem was largely solved by the government for those growers (the bulk of them in the Eastern Highlands) who lived close to Goroka. The airfield at Goroka was in airmiles the closest of the highlands airstrips to the ports of Lae and Madang. Because of wartime needs it was the best developed in the highlands, and the growers were able to back-load at low rates on DC-3 aircraft which brought in materials and supplies for the government. For surface transport the growers could use the 'crude but extensive network of roads laid down by the administration' using Gorokan labour. [123] By 1964-1965 the road network included, however, a highway to the sea, obviating reliance on air transport. In addition, HFSA and the Western Highlands Chamber of Commerce established a Joint Transport Committee in 1968 to present the opinions of planters and businessmen on road transport to the administration. The committee also presented a submission to the World Bank Mission of 1968, putting the case for improved access to the coast. [124] There is no doubt that by this time the planters were well entrenched, and together with the kiaps formed a small tightly knit community the aspirations of which were aptly expressed by the mast-head of the HFSA 'New Guinea Highlands Bulletin': 'We are Here to Stay'.

The impact of the planters

The opening of the highlands for land alienation, between May 1952 and 1954, had a considerable impact on Papua New Guineans as suppliers of land [125] and of labour for the new plantations, but it also quickly affected some of them as potential cash crop producers in their own right. Certain planters and administrators turned planters (e.g. Taylor, Downs) viewed their position as extending beyond short term individual profit making and involving additionally 'aiding Gorokans to enter the money economy'. Believing that they had uniquely good relations with the villagers of the Eastern Highlands, such planters set out to promote production for export among Papua New Guineans. They constructed an impressive ideology involving the plantation as a model to smooth the way, 'partnership', with supposed references to policies of the time in the Central African Federation and to the land reforms and other consequences of the outbreak of peasant rebellion in Kenya. [126]

It was clear, said Finney, 'that European plantations did in fact serve as demonstration plantations that stimulated Gorokans ... to emulate European production techniques'. [127] Rapid results were, however, achieved via the direct planter/patron and trusted Papua New Guinean worker/client relationship. Finney referred to Jim Taylor's relationship with Biato Hero (sometimes Khasawaho Biato Heiro), a policeman during the war, then Taylor's personal servant, and subsequently foreman on his plantation. After the plantation was under way, Taylor gave Biato seedlings, helped him to plant them on Biato's land, and encouraged Biato to tend them. The outcome was that Biato became 'one of the first major New Guinean coffee growers' with, as early as 1957, some 5,000 trees and an annual income in consequence of $4,000. [128] In 1977 he owned six coffee gardens and a small factory near Goroka, employed seven wage labourers and had an income of about K5,000. [129]

'A few' foreign planters, according to Finney, proffered this kind of direct patronage, [130] and few Papua New Guinean coffee growers did as well as Biato (though some did even better). The kiaps and planters needed to preserve good working relations with the highlanders. The land that they occupied was usually theirs because they had personally negotiated its sale with the land owners or their representatives. Y. Baron Goto (see above) pointed out in his 1952 report that most of the plantations were badly under-capitalised, and the capital-short planters still needed to resolve the problem of labour. Perhaps it is significant that Biato had been the bosboi of Taylor's labour line.

If the planters gained land and labour from the 'partnership', then a few Papua New Guineans gained access to technique and technology. The production techniques which the plantations (the efficient ones, that is) displayed were however, of a comprehensive kind. Processing techniques involved the use of equipment such as pulpers and driers which were essential to efficient production, but only those aspirant indigenous coffee growers who could gain access to the machinery, by direct and costly purchase or by hire, could profit from the demonstration. The planters quickly established a transport and marketing infrastructure without which no coffee cherry could be sold by a farmer and no beans moved to Madang or Lae and exported. The use of these facilities was again on a commercial basis. Papua New Guineans who profited from the system were those who not only saw and understood, as it were, the need for trucks and buyers, but could themselves purchase trucks and establish processing and marketing facilities. Acquisitions of this kind were naturally on a sharply differential basis.

Some foreign planters also went beyond the simple 'partnership' stage in their relations with indigenous growers. The new role was largely a development of the 1960s and early 1970s and, while it was not a matter for public discourse such as the vaunted 'partnership' had been, it deftly focused on the small number of local growers who had by then distinguished themselves as more efficient and successful than their fellows. Influenced by a further interpretation of colonial Kenya (where the Swynnerton tenure conversion plan had been designed to promote an African rural bourgeoisie as barrier against peasant revolution), some foreign planters in the Eastern Highlands aimed to promote successful indigenous growers as bulwarks of the plantation system. The Highlands Farmers and Settlers Association sought to recruit prominent national growers so that they might 'identify with European growers' and oppose any disruption of the coffee industry through the break up of the plantations. They also advocated that credit be advanced to promising Papua New Guineans so that they could 'buy up some of the smaller European plantations' and 'truly become part' of the 'European' plantation system. [131] Sinake Giregire was a rising Eastern Highlands farmer assisted in this way, who became a member of the House of Assembly from 1964-77, a ministerial member under colonialism, and leader of the Country Party. He gratefully said, 'Plantations, which have been criticised, have done a lot of hard work. I do not want Papuans and New Guineans to oppose the planters'. [132]

The plantation system had thus a firm impact on pre-capitalist systems of production, appropriating land and labour, and assisting in turning Papua New Guinean agriculture towards cash crop production for export. Its influence was elitist both economically and politically, but this was not out of keeping with the hierarchical tendencies in the old societies. It prepared the ground for the establishment of an indigenous rural capitalism. Because the plantations existed, said the chief agricultural officer in Goroka, 1954-55, with greater prescience than he probably intended, the people 'knew that there was something at the end of ... the rainbow'. [133]

DASF and the growth of coffee production

In 1937 regular patrols were being made through the Goroka area, and between 1937 and 1941 the luluai system was introduced. A report by Armistead late in 1945 stated that most villages in the environs of Goroka had between two and four luluais, and that most villages had Tok Pisin interpreters. [134] The luluai system was of prime importance, not only in the introduction of cash crops, but in the sale of land. In 1948 ten government chosen luluais representing seven land owning groups were asked by district officer Jim Taylor if they could negotiate with each other and with patrol officer William Sippo for the sale of a block of land. The precise boundaries were to be determined by the luluais themselves, and they received payment to distribute to their clansmen as they saw fit. The practice was, as Sack pointed out, [135] common in German New Guinea. The area totalled 250 hectares, and the price paid was 5 pounds per hectare. The district officer noted in June 1948 'it is almost high quality farming land and supports what is for New Guinea a heavy population. On an open market it could bring from 10 pounds to 50 pounds per hectare'. [136] The 154 hectares comprised more than half the land on which the present township of Goroka stands.

In 1953, systematic and regular DASF coffee extension patrols got under way in the Eastern Highlands, and in 1955-1956 a farmer training scheme commenced in Goroka based on a similar scheme started at Kerevat in 1932. Extension policy in the Eastern Highlands was to encourage Papua New Guineans to accept coffee as their principal cash crop, and it aimed at regularizing and expanding the size of holdings. The 'basis' of policy was to promote 'minimum-sized plots of one acre properly spaced and using correct shade technique', and extension workers were 'to give no encouragement' to the planting of smaller or irregularly laid out plots. DASF also aimed to establish correctly planted demonstration gardens at all sub-district stations and police posts. [137] The area under coffee plantings 'increased rather slowly' until early 1954, [138] but was followed by a rapid rise in the expansion of indigenous coffee holdings towards the end of the decade. By the middle of 1959, more than eight times as much coffee was planted in the Goroka and Kainantu sub-districts than had been planted prior to 1955.

Table 4.2
Known acreage of native coffee, Goroka and Kainantu sub-districts 1955-1959

	Goroka	Kainantu	Total
Up to 1955	158	60	218
1955-1956	131	47	178
1956-1957	45	60	105
1957-1958	301	30	331
1958-1959	832	155	987

Source: Letter from DAO D.J. Kingston to Highland Farmers and Settlers Association, November 29, 1959; File No. 1-6-A(1), DPI Archives, Goroka.

Just as kiaps and didimen (agricultural extension officers) had played a crucial role in the development of agricultural capitalism in the district on one level, their counterparts among the colonised, the luluais and agricultural assistants, played an equally crucial role on another level.

Because coffee seed retains its viability for only a short period it was first necessary to aid its diffusion by building coffee nurseries in the villages. These nurseries were planted close to the rest house established in the village for the kiaps and didimen and were left under the care of the village luluai. In July 1953, the extension officer for the Eastern Highlands district, R.W. Cottle, reported that nine nurseries had been established in the Goroka district. Of these nine, six were being looked after by luluais, and three by 'native men'. Two nurseries established in Simbu sub-district and one at the Henganofi patrol post were being looked after by the respective kiaps. It was known that 'some headmen took a large share of the seedlings grown in the communal nurseries for planting in their own plantations'. [139] In addition, it was quite common for such men to use clansmen as 'essentially unpaid workers to clear and plant their plantations'. [140]

One of the six luluais was Soso Subi (sometimes Souso), who was by 1950, said Finney, 'also established as a big-man in the traditional sense'. Soso was 'one of the first Gorokans to seek out Robert Cottle for advice on coffee when the latter arrived in Goroka in 1952'. And 'Cottle jumped at the chance', for 'Soso seemed to be able to command the necessary land and labour'. [141] Thereafter,

> under Cottle's direction Soso and his clansmen cleared and planted a large nursery and prepared ground for transplanting the seedlings the following year. When the seedlings were ready for transplanting, Soso did not hesitate to take a lion's share for his own use or to employ his followers as unpaid labourers to help him plant his personal coffee plot, the nucleus of what was to develop as one of the first of the large Gorokan-owned plantations. [142]

Such were the special qualities and leading position of those whom Cottle and Finney referred to merely as 'interested men', and who were enormously advantaged by DASF's willingness to establish nurseries, help them select proper land, clear it, and then plant coffee seedlings and shade trees on it. [143] Brookfield noted that in the wet season of 1953-1954 in Simbu, those who were encouraged by the government to plant coffee were 'a selected group of leading men'. By the early 1960s a few men had obtained as much as a hectare of coffee, and by then such innovators had 'the smell of money keenly in their nostrils'. [144]

In addition to the head start taken by luluais, those Eastern Highlanders who came into early contact with DASF as employees were encouraged to take seeds and seedlings to their villages and put them and their new expertise to good use. When Finney conducted his study of Gorokan entrepreneurs in 1967, he found that of the ten big businessmen that he identified, two had been luluais, two had been workers on the Aiyura agricultural station and one had been an agricultural assistant working with DASF extension patrols. The ten cultivated an average of 9,000 coffee trees, ranging from 4,000 to 24,000. The average grower around Goroka at that time cultivated only some 450 trees.

Although the coffee nurseries were said to be 'communal' by DASF extension staff, the didimen took care to emphasise that the ownership of coffee trees must be on an individual basis. With the beginning of coffee production in the early 1950s, said Barrie, 'land tenure problems were settled before coffee areas were planted'. Thereafter the policy was 'to encourage individual, or at the most the small family unit, ownership of coffee blocks'. Communal coffee nurseries were encouraged, but simply as 'a labour saving expedient and to facilitate close supervision and instruction' by DASF. [145]

This individuation was compounded by the tendency for extension officers 'to direct their efforts to the few large growers who more closely met their western models of what a proper small-holder plantation should be like, in the hope that these 'better' growers would be examples to others'. [146] The policy of helping out the big growers soon had the obvious effect of making the big growers very big. DASF policy was to have every grower cultivate a minimum of 300 trees, but this was seldom achieved by more than half the growers. The following table is derived from a coffee census conducted by lands officer, R.J. Giddings, in the lower Bena area prior to payment of compensation for coffee trees standing on a block of land to be purchased by the government. It is probably then quite accurate.

Table 4.3
Coffee trees and coffee growers of the Ketarabo Clan and Upegu Clan, July 1965

	Ketarabo Clan	Upegu Clan
Number of growers censused	37	19
Number of growers possessing less than 300 trees	18	12
Number of trees owned by the largest grower	980	1,206
Number of trees owned by the smallest grower	9	94
Average number of trees per grower	244	428

Table 4.3 demonstrates that significant differences within 'clans', consistent with pre-capitalist inequalities, had emerged in scarcely more than ten years of state activity. The significance of the figures is not only that the biggest growers possessed 3-4 times more trees than the average grower, but that the attempt to preserve a minimum number of trees failed so badly so early, with half to three-quarters of the growers in these groups not meeting the 300 target. Table 4.4 below outlines the average number of trees per grower in 1965-66 by census division. The mean of these means stands at 192 trees per grower, suggesting that in the Eastern Highlands as a whole, the target of a minimum of 300 trees per grower was well short of being achieved. The policy of 'helping the big growers' was frequently a policy of helping the big men since the big man had the necessary control over land and labour to make himself a big grower in the first place. Following a study of eight highlands villages in 1975, Harris noted that 'big men were the first to begin coffee plantings in their villages ... the pattern that emerges is for big men to plant coffee and to be followed, between two and three years later, by a wave of plantings by ordinary men'. [147]

The policy of helping out the larger growers was also a <u>de facto</u> policy of helping those to whom the <u>didimen</u> had easiest access. Early plantings were concentrated around the major centres of Goroka, Kainantu and Kundiawa, and along the main road. [148] As Table 4.4 below indicates only growers in the Goroka sub-district had reached the 300 average by 1965-66. In particular, the Lowa census division, in which Goroka town is situated, had by far the largest number of growers and the largest average number of trees per grower. The DASF approach to agricultural development increased inequalities not only between individuals, but also between areas.

Table 4.4
Number of growers and average number of trees per grower, Eastern Highlands census districts, 1965-66

Sub-District	Census District*	Number of Growers	Average Trees per Grower	Sub-District Average
Goroka	Asaro	2,704	312	346
	Bena	3,022	285	
	Lowa	28,989	441	
Kainantu	Agarabi	1,668	103	138
	Kamano	1,439	173	
	Gadsup	1,217	138	
Henganofi	Dunantina	1,926	25	25
	Kafe	1,700	25	
Okapa	Gimi	1,201	194	199
	Keingani Kaniti	1,439	226	
	South Fore	981	259	
	North Fore	866	277	
	Auyana	793	37	

Source: Goroka Sub-District Annual Extension Report, Crop Statistics 1965-66.
Kainantu Sub-District Annual Extension Report, Crop Statistics 1965-66.
Henganofi Sub-District Annual Extension Report, Crop Statistics 1965-66.
Okapa Sub-District Annual Extension Report, Crop Statistics 1965-66.
DPI Archives, Goroka.
* Excluding Unggai, Yagaria, Tairoa, Fayantina for which information was either insufficient or clearly inaccurate.

The previously quiescent Australian state had been strongly reminded of the strategic importance of Papua New Guinea by the Pacific war. The post-war 'second boom' in Australia and the high price obtainable for coffee due to the Korean War also helped stimulate renewed interest. In 1946-47 the Australian government committed to Papua New Guinea more than forty two times more money than it had done in 1940, and about eight times more than it had spent in 1945-46. [149]

A direct effect of the renewed interest of the colonial state was the creation by DASF in Papua New Guinea of a Division of Agricultural Extension, employing 54 agricultural officers, which in the late fifties translated itself on the ground into a rapid increase in DASF field staff in the Eastern Highlands, as the following table demonstrates.

Table 4.5
DASF extension personnel, Eastern Highlands district,* 1958-60

	1958 June	1958 December	1959 December	1960 December
Agricultural Officer**	5	8	8***	17***
Agricultural Fieldworkers	4	4	4	8
Fieldworker Trainees	27	46	Not recorded	19
Farmer Trainees	-	-	-	68

Source: Six Monthly Reports 'Advancement of Native Agriculture', File 38-5, DPI Archives, Goroka.
* Includes Simbu Sub-District
** Includes officers on leave
*** Includes designation 'Project Manager'.

This rapidly growing cadre-force implemented DASF's policies towards land ownership and coffee production which were major catalysts of rural class formation. By the 1960s, the class nature of the government aid programmes was increasingly evident. Assistance for coffee was already concentrated on relatively large scale production in a few villages, and the colonial state became concerned with the development of individual land tenure and the production of 'a small class of purposeful elite farmers capable of responding to the opportunities opened up to them'. [150] The melding together of highlands pre-capitalist social relations with the new forces of production introduced by the colonial state and the planters, cemented by a firm policy of 'helping the man who helps himself', led to the deepening of already existing social inequalities.

Speaking to the House of Representatives in Canberra in 1962 Hasluck noted, '... the big majority of the present population [of PNG] are likely to find their advancement in changing from village subsistence to cash cropping, forming a native peasantry that ... will not be a major employer of wage earning labour'. [151] In so speaking, Hasluck was not announcing new policy, but was projecting into the future what had been occurring at an accelerating rate over the previous seven years. He was wrong in only one respect. Before the next decade was out, a small number of the new peasants would be quite significant employers of 'wage earning labour'.

Cattle enclosures and tenure conversion

As early as 1956 concern was being expressed by DASF at the rapidity of the spread of 'native coffee'. In April of that year Henderson, the acting Director of Agriculture, wrote to the district commissioner in Goroka that 'the policy should be to continue active coffee extension work in those areas where the crop already is planted and to concentrate in such areas on the improvement of technique and the bringing of existing plots up to optimum size, but not to further encourage the planting of the crop outside these areas'. [152] By 1959 orders had been issued to DASF extension staff to reduce the emphasis on coffee planting, because of the 'conflict of interest between indigenous and expatriate growers'. [153] In 1961, Australia (and hence Papua New Guinea) became a member of the International Coffee Agreement Organisation (ICAO). The terms of membership proscribed the further alienation of land for European plantations, and extension work designed to spread coffee growing among the indigenes. The joining of ICAO meant the end of the active phase of coffee extension work [154] and not until the late 1970s was expansion to be so enthusiastically promoted by the state again (see chapter 6 below).

The end of the phase was to have a dramatic effect on the uneveness of development between sub-districts in the Eastern Highlands. The Baruya of the Wonenara sub-district, for example, 'discovered' by Jim Sinclair in 1951 and settled among by the government only in 1960, were supplying labour for Gorokan and coastal plantations in 1963, but did not receive cash crops until 1973-1975. [155] The isolation and underdevelopment of this part of the Eastern Highlands is strikingly apparent in terms of infrastructure and education, [156] as well as in other basic economic indicators.

With the decision to stop actively creating more coffee growers, DASF turned its attention to other forms of agriculture such as passionfruit, peanuts, vegetables, pigs, and particularly cattle. The change away from coffee was assisted by the announcement in 1959 by the Minister of Territories, Paul Hasluck, that agricultural extension was to be stepped up, and that it was to involve 'close supervision of individual farmers'. DASF accordingly adopted the 'trickle-down' theory of diffusion of innovations which involved paying almost exclusive attention to 'progressive farmers', thus legitimising and accelerating the tendency of its earlier work. The change away from coffee was also assisted by a slump in coffee prices in 1961, and by the desire of the government to offset adverse effects of this drop on the balance of payments. In 1959-60 Territory imports of fresh, chilled or frozen beef (excluding tinned meat) amounted to the equivalent of 2,100 head [157] and the government decided to reduce this capital outflow. This choice was facilitated by the existence already in the colony of a herd which had grown from 3,722 head in 1951-52 to 17,000 in 1961. [158] But the unsupervised growth of the national herd worried DASF officials. Cattle obtained from missionaries and white cattle owners, had been owned by indigenes at least since 1958, and by the end of 1960 indigenous owned cattle totalled 288. DASF were concerned to control the development of the cattle industry among indigenes at least to minimise the problem of disease.

In 1960, DASF established pilot cattle projects at Korofeigu, Safaiyufa and Kapagumarigi in the Bena census division. The controlled introduction of livestock into the Eastern Highlands was based initially on the supply of cattle from DASF's Goroka piggery and overland from the Baiyer River in the Western Highlands until in 1965 the Korofeigu livestock station, located half an hour's drive east of Goroka along the Highlands Highway, was opened. Bena was selected because of its rolling grasslands and the people to whom materials were provided by the colonial state and to whom cattle were sold at 15 pounds per heifer and 30 pounds per cow, were carefully selected by DASF field officers. In Korofeigu 'a very powerful village group' which had in the past 'dominated much of the Bena region' was chosen. This village owned 'huge areas of land most of which [was] open grassland showing great potential for cattle'. In addition, 'their early planting of coffee [had] made these people a very rich group amongst the natives'. [159] Of the six men who controlled the project, one had been a <u>tultul</u>, one was a councillor and two were agricultural committeemen. As for the people of Kapagumarigi, 'the people of this village together with the neighbouring Korofeigu's [were] a very powerful group amongst the Bena people, and they were thus selected for one of the pilot cattle projects ...'. In Safaiyufa, it was individual differentiation rather than group differences which mattered; 'the majority of the cattle [were] owned by one Luki who [was] employed by the DASF'. [160]

The basic policy towards the development of cattle farming and cattle ownership was spelled out by DASF director, Henderson, in 1960.

> Considerable thought has been given to the best method of introducing cattle into the native economy ... The Department's experience with co-operative ownership of the means of production, as distinct from marketing, through the Territory has not been very happy ... in a large number of cases co-operative ownership ... just stultifies future development. It appears to me that with cattle the advantages of private ownership and co-operative operation can be successfully combined. Our plan envisages the individual ownership of animals with co-operative management ... I do not believe that progressive natives or a progressive family would have any difficulty in raising the price for one beast ... In these trial runs we propose to carry out in your district we are seeking the participation of the more advanced and progressive people who would have sufficient financial standing to provide the money required. [161]

The plan was to establish carefully supervised cattle projects in the district among the 'progressive' farmers. A progressive farmer was generally one who had succeeded in coffee, and therefore was or had been (as shown) a big man, <u>luluai</u>, <u>tultul</u>, councillor or DASF employee. The scheme also involved fencing a paddock, in which the cattle would be placed at night, while during the day the cattle would range freely under the direction of a cattle herder. The night holding paddock was to be at least 2 hectares, planted with improved pasture and containing a suitable set of yards. The timber for yards and fencing cost between 40 and 120 pounds. [162] It is clear that in order to become a cattle farmer, a villager needed to be very progressive indeed. He had to raise the money to buy the stock, to build fences and yards, and he had to command enough labour to help him carry out the fencing. As was later noted by McKillop, 'In the Highlands areas the only farmers likely to have enough capital were the 'big men' who planted up large areas of coffee. The majority of the population were excluded, either by lack of capital or lack of land suitable for cattle grazing'. [163]

DASF had not abandoned the cause of promoting a peasant based commercial agriculture, but had merely shifted emphasis from coffee to cattle. This was supported by the views of the World Bank mission which toured PNG in 1963 and submitted its report in September 1964. It recommended among other things, an increase in primary production, specifically a tenfold increase in cattle numbers within ten years, and the 'principle' of 'concentration of effort', of expenditure and manpower, in areas and on activities where the prospective return was highest. [164] This development principle was already at work in the production of coffee.

The number of cattle projects and the number of cattle on them grew rapidly. In November 1965, it was obvious that demand for live cattle was outstripping supply, a fact noted by the district veterinary officer in the Eastern Highlands in a report to the Division of Animal Husbandry. He added that '[I]n the Eastern Highlands 15,000 pounds is held in deposit for cattle development. In the last three months in Goroka alone, deposits have totalled 2,000 pounds per month'. From 1963-64 to 1965-66 the number of cattle in the district more than doubled, as did the number of planned projects. The number of stocked projects trebled.

Table 4.6
Cattle projects and cattle numbers, Eastern Highlands district * 1963 to 1966

	1963-64	1964-65	1965-66
Total number of cattle	332	493	856
Total number of stocked projects	34	53	110
Total number of unstocked projects	31	87	73

Source: McKillop, EHD Cattle Projects Annual Report 1965/66
DPI Archives, Goroka.
* Including Simbu Sub District.

The speed at which cattle were being taken up by Eastern Highlanders was not matched by the evenness of their dispersal. After just over three years of extension work on cattle it was already clear that, as with coffee, significant differences were emerging between areas and between individuals in the ownership of cattle which was hardly surprising for as McKillop reported in 1965, 'in the great majority of cases the money to buy fencing and cattle is derived from coffee'. [165] This official also noted with some satisfaction that although one project was permitted to be built up with many owners (each individual owning at least one beast), in practice, 55 per cent of the projects were financed by one or two men. [166] Since it required 4 heifers to begin a project at K60-K70 each, plus the money for fencing and pasture improvement, the establishment of a cattle project by one or two men represented considerable individual control of capital.

Having successfully established a class of big growers, DASF was well on the way to establishing a class of relatively wealthy mixed farmers. This tendency accelerated through the 1960s and into the 1970s. The cattle projects, particularly the larger ones, also tended to cluster around the developed Kainantu-Goroka corridor leaving the southern part of the region largely untouched. While this was partly attributable to geography - the rolling grasslands of Bena and Kainantu being particularly suitable - it is also the case that the diffusion of cattle followed the same social and geographical path as coffee and sweet potato before it.

Table 4.7
The number and size of cattle projects in the Eastern Highlands 1963-64

Number of cattle per project	Number of projects in each Sub-District		
	Goroka	Kainantu	Okapa
0 - 5	5	3	2
6 - 10	6	4	1
11 - 15	4	1	
16 - 20	2	2	
21 - 25			
26 - 30	1		
Total number of Projects	18	10	3
Average number of cattle per project	10	8	3

Source: Native Cattle Project Annual Report 1963-64 DPI Archives, Goroka.

The vision of the 'selector', the 'close settler' of Australia's domestic history, had for a long time enthralled colonial administrators and agricultural policy makers in Papua New Guinea. In the early 1960s it came closest to being realised. In August 1964, R.N. Amos, the district agricultural officer, stated, 'The ultimate aim has always been to have individually owned stock on individually owned land and current projects are only regarded as temporary holding paddocks for individually owned stock ...' [167] For McKillop, the concrete realisation of this vision was 'the development of small mixed farms of 10-20 acres with closely integrated cash cropping, food gardens and cattle on improved pastures'. The problem was one of the availability and ownership of land.

As early as 1954 Hasluck had suggested that 'in the future, the needs of an expanding native population may not be met by inheritance in the traditional system of tenure and we may have to keep in mind the possibility that some of the natives themselves may require land under a European form of tenure'. The policy was reiterated by him in April 1960. He announced that the ultimate and long term objective of the government was to introduce a single system of land holding regulated by the central government which would provide for secure individually registered titles. The necessity for this change was argued on the premise that since agriculture was the basis of the colony's economy, changes in land tenure were necessary to ensure economic progress. [168]

Legislation to give effect to this policy was subsequently passed and the following acts came into effect in 1963 and 1964: the Land Titles Commission Ordinance 1962, the Lands Registration (Communally Owned Land) Ordinance 1962, and the Land (Tenure Conversion) Ordinance 1963. Under the first, the commission had the function of hearing land claims and disputes. Under the second, customary land, the boundaries and ownership of which had been determined by the commission could be registered as communal land in a special register. The third ordinance provided for the conversion of customary land into individual freehold tenure, and it was this which was to be the principal legal instrument in the transformation of the agronomy.

Tenure conversion enabled applicants, if long established usufruct could be proved and if no other claimant should come forward to veto the application, to gain full individual freehold or a 'fee simple' title. Much therefore depended upon an individual's personal standing and reputation and his relations with kin and those in the vicinity of the land in question. It is noteworthy that the 1973 Commission of Inquiry into Land Matters referred to the opportunities thereby given to a 'single ambitious and forceful person'. [169] John Akunai (sometimes Rove or Rovelie) was the first successful applicant, in May 1967, for land tenure conversion in the Eastern Highlands, and Sinake Giregire, in December 1967, was the second; Akunai was granted title to 9.19 hectares at Okiyufa and Giregire obtained 23.58 hectares at Asaro. Both men were large scale coffee growers.

Between May 1967 and December 1973, there were fifteen successful tenure conversion applications in the Eastern Highlands, in some of which a relatively small monetary compensation was made to the former rights' holders, and where in most cases the applicant's social and economic status was crucial to the outcome. Sinake Giregire in July 1971, obtained freehold over another 1.9 hectares at West Goroka where, it was made clear at the hearing before the Land Title Commissioner, Giregire had no blood relationship with the so-called traditional owners, but where Giregire explained that he had already erected two subtantial concrete buildings on the land concerning which (in the words of the Senior Commissioner) he 'intended forming a private company in which the former right holders of the land would be offered shares'. [170]

When John Pokia got freehold over 47.49 hectares in the Okapa sub-district in 1971, no compensation of any kind was sought, and one witness declared that Pokia would 'look after us'; the Commissioner related his decision to the fact that the applicant was 'an enterprising young man' who was president of the Okapa local government council, a health education officer, and trade store owner; and Pokia soon displayed his enterprise by becoming a member of the House of Assembly in 1972. The decisive importance of success in commercial farming, politics, and trade was also clear when Kege Yasinamo obtained freehold over 20 hectares at Okapa in October 1971; Yasinamo held a sawmill on the land which brought him some $4,800 per annum, he was a councillor, cattle and trade store owner and, in the Commissioner's words, 'one of the most progressively minded men in his area'. [171]

'Tradition', Gusfield has said, 'is not something waiting out there, always over one's shoulder. It is rather plucked, created and shaped to present needs and aspirations in a given historical situation'. [172] The manipulation of the control over land in pre-capitalist PNG was, as already described, largely in the hands of the big man, who could take greatest advantage from the existence of 'some mechanisms for permanent transfer of rights' in the old tenure systems. [173]

The changing and highly malleable nature of so-called tradition and custom in the hands of such men is indicated by Giregire's position at the end of 1978. He had lost his long time parliamentary seat at the elections in 1977 and had seen his Country Party evaporate, but now had more time for business. He then produced, he said, some forty tonnes of coffee per year and claimed an income after tax of K9,000 in 1977. He had also 12,000 cabbages planted and had plans for cattle, but nonetheless stressed that he managed his land in accordance with custom. He contributed to people's social needs like the expenses associated with funerals, he encouraged others to start businesses, and he lent money for such things as the purchase of a car; he had in fact some

K10,000 owing to him from money lent in this fashion at the time. He also stressed his general economic contribution to the area, an access road, electricity, extension, etc., and said that he allowed people to garden on his land, but he eventually intended nevertheless to subdivide his land for individual holdings. He was still in the process first mentioned at the Lands Commission hearing in 1971, of forming a company but said that dividends would be 'paid this Christmas'. Giregire's commercial position was therefore far from weak, but his supposed customary position was characterised by the large indebtedness he had encouraged among local people and perhaps by plans for a system of tenant farming. [174] Tenure conversion, on the other hand, had clearly provided what he and other applicants had sought security over land, availability of bank loans, and future succession to their children.

The problem then confronting DASF was that farmers were faced with an extremely long wait while respresentatives of the Lands Commission investigated their claims. By April 1971, a total of 427 conversion orders had been received by the Registrar General from the Land Titles Commission and 273 individual titles had been registered. Only twenty one of these were outside the Northern District where a pilot project had been mounted based on the distribution of clan land to members of the clan. [175]

It was officially suggested that 'in more sparsely populated areas where land is available, its alienation and development would serve a similar role to land tenure conversion'. [176] Shortly after submission of this report, administration staff in the Eastern Highlands began considering a suitable block of about 1,200 hectares, which had originally been bought for tea, in the Kainantu sub-district near Aiyura. The Arona cattle project eventually comprised 15 blocks ranging from 72 to 132 hectares, which were specifically intended in DASF's official view, 'for people who have had a lot of experience in rearing cattle and running a business'. DASF called for applicants to buy and settle on the blocks and submitted a list of 35 names ranked in order of credit worthiness and farming ability to a selection committee. The selections were made in January 1970, with seven of the fifteen blocks going to farmers from Goroka, though eligibility was not only district but nation wide. The regional rural development officer wrote to the assistant director, Development and Marketing, Port Moresby,

> Three blocks have been sacrificed as pawns, to politics. Two of them were granted to illiterate cohorts of a ministerial member, neither of whom can speak pidgin or had the experience needed or any intention of raising cattle. Another was granted to a 21 year old son of a local member [of the House of Assembly]. Four other blocks were granted to people who we consider very poor credit risks ... In effect 50 percent of the scheme has been sabotaged. [177]

The significance of the Arona allocations and this reaction is not so much to do with corruption, but with its implication for DASF. It was becoming clear that technical judgements and decisions were no longer sacred. Rich farmers and politicians, perhaps former clients of white patrons, realised that old systems of patronage had new roles to play. Papua New Guinea members of the House of Assembly were realising to some degree the essentially political nature of 'technical' decisions. The Arona scheme was the first and last large resettlement development in the Eastern Highlands.

DASF's aim of 'small mixed farms' and Hasluck's plan to subdivide

7,500 blocks by 1967 were wrecked by the intervention of the World Bank. The visit of the World Bank Mission to Papua New Guinea in 1962-63 and the subsequent appearance of its report in 1964 was to have quite an effect on agriculture in the Territory, at least insofar as it confirmed and by doing so, accelerated, existing tendencies within the new, slowly emerging, system of capitalist agricultural production.

It was noted as early as mid-1966 that 'one pleasing feature has been a trend toward fencing of larger areas ... Those projects of up to 100 acres appear to be well on the way to becoming complete pastoral holdings of an extensive nature'. [178] The DASF official did not perceive in this a tendency antagonistic to his vision of the small mixed farm. The World Bank however did, and it came down heavily on the side of the larger cattle farmer. The World Bank commented that farmers confined to small areas with small herds, combined with the policy of encouraging the conversion of individual usufructory rights over clan land to freehold title, tended 'to continue and expand the creation of potentially non-economic holdings'. [179] The World Bank suggested that the development of large areas of land for cattle farming as single undertakings would be advisable. By 1968-69, livestock officers with DASF were using the Animal Disease and Control Ordinance to destroy the remaining 'night paddock plus free grazing' system of animal husbandry, by insisting that stock be kept fully enclosed. [180]

By mid-1965 about 820 acres had been enclosed in the Eastern Highlands, and by 1971-72 this had risen to about 20,000 acres. Although the acreage had increased by a factor of 25, the number of stocked holdings had barely doubled. Table 4.8 below summarises comparative data for 1965 and 1971, [181] and Table 4.9 describes more fully the situation in the district in 1971.

Table 4.8
Cattle holdings, Eastern Highlands, mid-1965 and mid-1971

	1965	1971
Cattle population	856*	2,376
Number of stocked projects	110*	227
Area fenced (acres)	827	20,000***
Average area enclosed (acres)	7.5**	88
Average cattle per project	8	11

Source: R. McKillop, EHD Cattle Projects Annual Report 1965-66
A.W. Marsh, Native Owned Livestock Statistics, Highlands Region, June 30 1971. DPI Archives, Goroka.

* Including Simbu sub-district.

** Obtained by dividing the area fenced in the Eastern Highlands by the number of projects in the E.H. plus Simbu sub-district. The average size of the enclosures in the E.H. in 1965 was therefore larger than this figure.

*** Approximation based on Marsh.

As with coffee plantations, the size of the cattle farms varied enormously. The farms of less than 20 acres accounted for 45 per cent of the projects, but only 17 per cent of the cattle. The largest farms, of more than 400 acres, amounted to 4 per cent of the total, but carried 20 per cent of the stock. In a circular memorandum dated July 1965, D.A. Boniwell, regional agricultural officer, advised that future cattle projects would be developed only in those areas with resources capable of sustaining a herd of 25 head, thus reinforcing the concentration and centralisation of cattle farming.

By 1971, in less than a decade after the beginning of the DASF cattle programme, this centralisation and concentration had produced marked inequalities, as can be seen in Table 4.9 below. Thirteen per cent of the projects, covering at a conservative approximation 10,000 acres, contained 44 per cent of the cattle.

Table 4.9
A summary of livestock statistics for the Eastern Highlands, June 30, 1971

Fenced Area (acres)	Number of Projects	Number of Cattle	Average Number of Cattle per Project
Less than 20	102	409	4
21-100	55	10	68
101-200	40	60	215
201-400	22	418	19
More than 401	8	492	61

Source: A.W. Marsh, Native Owned Livestock Statistics, Highland Region, 30 June 1971. DPI Archives, Goroka.

The 1960s saw the systematic introduction of cattle into the Eastern Highlands and their eager adoption by the emerging peasantry. To an even greater degree than coffee production, more rapidly and more intensely, cattle farming became geographically centralised and socially concentrated. The tendency toward concentration of wealth, already existing in pre-capitalist societies had been shaped, adapted, but also tempered during the introduction of coffee. The emerging classes were further reinforced by state action during the enclosures of the 1960s, action which went beyond the introduction of cattle and the amending of land legislation.

The leap forward: the provision of credit by the state

The increase in the number of cattle holdings, and more spectacularly the increase in the size of the holdings, did not occur by DASF, or even World Bank, fiat. A major catalyst in facilitating the rapid growth of the cattle industry in the Eastern Highlands was the absence of credit. Land and labour were already available to a select few through older hierarchical mechanisms and infrastructure and marketing capability had been provided by the planters and by the agents of the colonial state.

The last major inhibitor to class formation and consolidation, the provision of credit, began to crumble in 1961. In that year pilot cattle projects were introduced, following the successful establishment of three in Bena in 1960. A further ten projects were established nationally, six in the Eastern Highlands, and financed through the Commonwealth Bank Rural Credits Development Fund. Impetus was also given to the easing of credit for indigenous farmers by the introduction of the Tenure Conversion Ordinance and resettlement schemes. Two credit organisations existed in the Territory which were prepared to lend money to indigenes, the Native Loans Board, established in 1955 and the Ex-Serviceman's Credit Scheme which was established in 1958.

The Ex-Serviceman's Credit Scheme made capital available to finance agricultural development by Australian and Papua New Guinean, ex-servicemen. The Ex-Serviceman's Credit Board required that land being developed be offered as security for the loan. Applicants were required to hold land under leasehold or freehold title so land held under traditional tenure was not acceptable as security. [182] The first loans to Papua New Guineans were granted toward the end of 1960, and when the scheme closed down in November 1962, 136 loans had been granted. [183]

The Native Loans Board was the main operating agency, and it was characterised by largely negative views and procedures. [184] The Fund's policy was that 'economic development rather than welfare should be emphasised', and such loans as it offered were 'intended to supplement not replace normal credit facilities'. 'Native land' was 'not suitable' as security, and 'only in isolated cases' would a loan be granted without security. Not only was a 'good title' to land necessary before it was acceptable as security, but the property also needed to be 'an economic unit from a European point of view and sufficiently attractive to be readily marketable in the case of repossession and sale'. [185] The processing of loans for cattle development was not surprisingly a 'long and tedious affair' in the view of DASF in the Eastern Highlands and their policy in consequence was that the Native Loans Fund 'should only be used as a last resort' for such as fencing loans. [186]

In 1962 the first savings and loans society was registered in the Territory and by March 1968 there were 458 savings and loans clubs and societies operating, lending a total of about $42,000 to indigenes. Savings and loans societies in the Eastern Highlands started after those on the Gazelle in New Britain, which was also an agriculturally advanced area, and in terms of numbers of clubs and total funds, the Eastern Highlands ranked third behind East New Britain and Morobe in 1968. The three districts between them held slightly less than two-thirds of the clubs and societies in PNG and 61 per cent of the funds by the end of March 1968. Of the three the Eastern Highlands was the smallest, retarded as the deputy registrar of Savings and Loans Societies remarked, by the problem of illiteracy. The Eastern Highlands was strong enough however to be with the Gazelle, the first area to form a district federation ('League') of savings and loans societies. [187]

As could be expected, given the overwhelming influence of emerging agricultural capitalism, 70 per cent of the nearly 4,000 loans totalling $414,000 granted by June 1968, had been for income producing activities. Significant in the national figures was the $16,900 loaned for the purchase of labour power, $13,400 for the purchase of land, $8,700 for the purchase of fencing wire and $7,100 for the purchase of agricultural equipment other than copra driers. [188]

During the 1960s and beyond the savings and loans societies remained important sources of finance for some Papua New Guineans. The total funds held by averaged out at $2,000 per individual society or club, thus limiting the size of the loans which was restricted to a maximum of one-fifth of the value of the deposits. Of money lent from 1962 to June 1968, the average loan stood at $103. [189]

The 1963 World Bank Mission reacted against these static and inflexible loans policies by making a strong recommendation that developmental credit should be made 'readily available ... in particular to finance small scale native agriculturalists'. [190] The leap forward in the state provision of credit occurred after 6 July 1967, with the opening of the PNG Development Bank which absorbed the two earlier instrumentalities described above.

The basic philosophy of the Development Bank as described by its first managing director, K.G. Crellin, was of 'lending to the man', and the Development Bank used five basic criteria for determining the credit- worthiness of the individual borrower: i) Is the proposition viable in relation to the cash and labour to be invested? Will the project return a reasonable income and allow the borrower to repay the loan? ii) Is the borrower assured of continued use of the land? iii) Has the borrower sufficient managerial skills for the projects? iv) What equity (cash, labour) can the borrower contribute? v) Is adequate supervision available? [191]

The Bank followed an open and flexible policy of lending on 'assured usage rights' to land, and a large volume of loan money soon flowed in the Eastern Highlands. In the brief period from 6 July until 11 October, 1967, $237,805 as allocated to seven projects, making the Eastern Highlands easily the leading district for the receipt of loans. From the opening of the Development Bank until 20 March, 1968, 13 loans totalling $473,997 were received and the Eastern Highlands was third among recipient districts; and for the period until 31 December, 1969, 66 loans were obtained totalling $1,240,000 and the district was fourth on the national list. [192] While the expansion of coffee growing was accorded rather a low priority by the bank, [193] and although a large proportion of lending in the early stages of the bank's operations went to foreigners, indigenous farmers in the Eastern Highlands were nevertheless well placed. During the ten month period from 1 July, 1970, loans worth a total of some $226,000 were approved for 122 Eastern Highlands' farmers, and from 1970 to 1975, $516,000 was loaned to 179 cattle farmers. [194]

This volume of credit reflected a high approval rate by the Development Bank, more than 80 per cent of all applications dealt with over the first nine months, but it also related to the 'quality of the work of our Agents in the field', as the bank gratefully acknowledged. [195] The Bank's agents were mainly DASF employees and those of the Eastern Highlands had obviously displayed a particular enthusiasm. [196] Active and successful collaboration between the established DASF and new Development Bank implied harmony of policy for the further concentration of key resources in few hands. DASF was the active agency of the bank in further accelerating the inequalities in indigenous agricultural production.

The Development Bank combined an appropriately flexible attitude to the question of land usage and control, [197] with the encouraging policy of 'lending to the man'. Thus those that were well placed to obtain loans included big coffee growers (the reluctance to lend for coffee excluded cases where credit was necessary to bring an existing plantation up to a so-called economic area), [198] local politicians, and trusty government employees. Each of the thirteen original block holders on the Arona cattle estate received a loan between $2-3,000 in the first year of operation, but one Ono Aia (sometimes Aiya), who was a local government councillor, also received a loan of $5,488in September 1972, which was increased to $7,588 in November. Ono Aia had obtained leasehold ownership of 573 acres at Arona. He held 45 cattle, owned some five acres of coffee, and became a nominated member of the Eastern Highlands interim provincial government. [199] Koroshuhu Masuwe of Goroka got ownership via tenure conversion of 73 hectares in January 1968, and at about the same time a Development Bank Loan of K24,100; and Norifo Kindino (or Palang/Palank), a long time employee of DASF, received a loan of $28,600 in 1969, one of the highest of its day to a national farmer, and soon after held 600 acres with cattle. [200]

While Development Bank officials were inclined to consider that loans of less than $500 were too costly to administer, they were extremely eager to lend for 'second stage development', even if the first loan had not been repaid fully. A cumulative process was strongly at work where success at various fields of agriculture, trade and politics were closely inter-related. Bank officials were as ready to push forward these tendencies as were didimen and a few planters were earlier. For the Bank's operations manager in 1971 the aim was explicitly one of rapid 'wealth acquisition' by a necessary few.

> [I]f the aim is to provide a man with economic wealth as quickly as possible, in our minds the only course to follow in an under-developed country approaching independence, then the man should be persuaded to undertake a forced saving, i.e. a loan obligation, and stick to the programme ... Really, there is a lot to be said for having a disciplined borrower perpetually in debt, providing such debt means that he is continually expanding in acquiring new wealth and that he himself is receiving some share of the increasing returns of that wealth to preserve his incentives and to lift his standards. [201]

These credit policies built firmly upon twenty years' encouragement of capitalist agriculture within the peasantry by DASF.

Supervision was the key to the success of the Development Bank lending schemes. It was provided by DASF when in 1967 its field officers were made agents of the Bank. The ready availability of credit encouraged the existing trend toward large holdings, which fitted in well with DASF emphasis on technique. For DASF larger projects meant more effective supervision, improved breeding performances and better disease control. [202] The new role of DASF extension officers placed a severe burden on existing staff, and accordingly, the number of DASF personnel in the district increased sharply from 20 in 1963 to 160 in 1973. [203]

The combination of credit and close supervision made possible through extra staff, led to a rapid rise in the number of indigenous-owned cattle in the country. The total rose from 379 in 1961 to 8,862 in 1970. By 1971, cattle loans to Papua New Guinean cattle farmers stood at $829,000. The loans extended to indigenes averaged about $1,600 in 1967-68 and $1,110 in 1970-71. The average cattle loan was less than $2,000 although two borrowers had received $50,000 and a few were in the region of $10,000. [204] The largest loan received by an Eastern Highlander up until 1970 was that of Norifo Kindino. Kindino who had worked 18 years with DASF, received particularly strong support from the organisations' livestock branch, and a loan of $28,600. In 1970 Kindino ran 24 head which he planned to increase by buying 80 breeder cattle from Australia, and 15 steers and two bulls from local sources. Over a six year period he planned to expand his herd to 170 head. With an expected turnoff of 50 beasts, he could anticipate an annual income of $7,000. [205]

Kindino's success story was perhaps atypical (and it ended rather abruptly). [206] Von Fleckenstein reported on the Ketarovo cattle scheme in Bena which commenced in 1969 and was a 'group' cattle project. He noted that:

> individuals did not receive benefits in proportion to the work and money they had put into the project ... When the money and labour put into the project is subtracted from the money received, it is evident that only the leaders and the old men ... have received any net benefits from the project. At the end of 1973, the others had mean net costs of K50 to K150 ... Many of the men who contributed some money and labour to the project ... felt that cattle was a good business and that the Ketarovo cattle project had been helpful in making Ketarovo well known, but they also felt vaguely uneasy that they were being exploited. [207]

The project leaders controlled the land, the Development Bank loan and the buying and selling of the cattle. The others possessed some cash from the selling of their coffee which they invested at a loss in the project, and their labour power which they gave free.

Class formation: the rise of a rich peasantry

The control of labour power was historically an aspect of the power of the big man, expressed within the village community particularly as control over young men and wives. At the early stages of a peasant based capitalist agriculture, therefore, the big men and the luluais who figured prominently made important use of the often unpaid labour of their clansmen (and sometimes mobilised villagers' savings too). The big man and the luluai Soso Subi, as noted above, used his clansmen and followers as unpaid labourers in the essential task of clearing and planting what became a substantial coffee holding; Bono Azinapfa, another big man and luluai, also established his plantation using his clansmen as unpaid labourers; and Bimai Noimbano, first an agricultural trainee and later a big man, called on 'traditionally recruited labour' in planting and exploiting coffee in a particularly impressive way. [208]While such newly established coffee planters may have made some return to their clansmen in extenuated and diffuse ways, the loans funds subsequently obtained by rising farmers certainly tended to sever the obligations of their formative years. [209] Sinake Giregire, a well established coffee farmer, was able in 1974 to build his Gire Coffee Factory with reportedly $70,000 in cash 'and a lot of friendly labour at no direct cost'. [210]

The labour of women and of children has been and still is important in cash cropping, and the usage is such as to go beyond the reliance upon family labour characteristic of peasant societies generally. [211] Apo Yeharigie, another of Goroka's early business leaders, had long associated with the first whites in the highlands, and he made use of the labour of his many wives, actually 'the backbone' of his labour force in what became highly successful market gardening and coffee growing activities. [212] The labour of women in general and of children remains important in coffee production, and a survey in 1976 indicated that in the Eastern and Western Highlands women carried out about 50 per cent of the harvesting work and children 15 per cent. [213] The exploitation of children's labour is a feature of the operations of some successful commercial farmers. Kangeto Yavise, for example, who had in 1978 some 7,000 coffee trees, and who was a member of the Henganofi local government council and of the Eastern Highlands interim provincial government, employed children from grades 1 to 6 at the local community school on his coffee farm. In return he paid their school fees of K4 per pupil each year and gave them some food. [214] The monetary gains to Yavise from this lopsided arrangement would have been considerable.

The use of wage labour had been occurring on a growing but little recognised scale since the early years of cash cropping. John Akunai, who had been a young interpreter with the early patrols in the highlands was one of the first to use paid labour when he began coffee production in 1955. In 1968, after being the first to obtain Tenure Conversion in the Eastern Highlands when he had a large market garden and owned 12,000 coffee trees with 5.5 hectares under full production, his wages' bill for permanent labour was an estimated $720 per year with an additional $1,000 for labour during harvesting. He was therefore a considerable employer of wage labour by local standards. He appeared before Rural Wages Boards to argue against increases in rural wages, and more recently his labourers numbered fifteen to twenty in the season. [215] Anderson's 1976 survey indicated the growth in the use of wage labour when he found that one-third of those inteviewed used paid labour on their coffee farms, and he believed that the ability to mobilise labour was 'of utmost importance' in the higher management performance of larger growers. [216] Wage labour was therefore not only becoming increasingly common among highlands' coffee growers, but its usage was a particular characteristic of the successful indigenous grower. It is at the basis of the coffee consolidation programme introduced at the end of the 1970s (see chapter 6 below).

As in Tanzania in the 1920s and 1930s, [217] the introduction of commercial agriculture for a regional or international market offers new opportunities for inequalities within indigenous societies, based on the ownership of property and the employment of wage labour, and resulting in the growth of a distinct rich peasant class with certain specific characteristics. In coffee production the key property acquisitions are land, trees and processing equipment such as pulpers. In the later 1960s, the ten Gorokan business leaders surveyed by Finney owned an average of 9,000 coffee trees with Sinake Giregire, John Akunai, Apo Yeharigie and Soso Subi all appearing in the first five places on assets held. [218] For the whole class of rich peasants more recently, the base line of property at least would be those holding in excess of 1,000 trees. Anderson in 1976 found that over half the growers surveyedin the Eastern and Western Highlands and Simbu owned less than 500 mature trees, and he estimated that only 6 per cent of growers held more than 1,000 trees. [219] Wage labour was used by as many as one-third of the growers surveyed, but the ability to mobilise labour was a vital factor in the higher performance of larger growers, and it was a particular characteristic of the most successful national farmers where it was both cause and consequence of the large size of their holdings. [220]

Ownership or access to processing equipment means that the coffee farmer avoids drudgery, obtains a higher price for his product and perhaps gains an additional source of income through the hiring of such equipment as pulpers to non-owners. In the early 1970s the price of a hand powered pulper was about equal to the average grower's annual income, and in the late 1970s the price was some K175. Collective ownership, often with more than five other growers was the norm. Anderson found only 11 per cent of his sample of growers owned their own pulpers, and that generally ownership became more individualised the larger the holding. Hiring charges varied by the locality, and in terms of a flat rate per bag or per season or day, proved 'extremely profitable', in Anderson's view, to the owner. [221] It is then a further determinant of class position within the peasantry.

In 1973 the total number of coffee growers in the Eastern Highlands was 53,700 and the total area cultivated was 7,992 hectares. In 1976, however, the number of growers had increased by 115 per cent to total 115,887, while the area cultivated had grown by only 47 per cent to reach 11,760 hectares. [222] The estimate that 6 per cent of coffee growers in the highlands owned more than 1,000 trees may be used to gauge the size of the rich peasantry. On this basis the number of rich peasants in the Eastern Highlands in 1976 as a proportion of the total Eastern Highlands' population was less than 3 per cent and was less than 5 per cent of the province's economically active population. [223]

While the comparative wealth and the political power of this very small group is literally outstanding, the characteristics of the other classes of the peasantry, a middle and poor peasantry for example, are less clearly manifest. Consideration of the position of the average grower, however, not only further clarifies the dominance of the rich peasants, but shows also that the situation of the mass of coffee farmers in the Eastern Highlands was changing for the worse. When Finney's leading rich peasants, in the later 1960s, averaged 9,000 trees between them, the general average in the Goroka area was 450. [224] By 1973, average tree numbers had declined to 359, the average coffee holding in the Eastern Highlands was about 0.15 hectares and average production was 152 kilos per grower. [225] In 1976, however, with the number of growers having increased much more rapidly than the area under cultivation, the average holding in the Eastern Highlands had fallen to 0.1 hectare, with just 250 trees, producing only 74.2 kilos per grower. [226] The very rapid increase in the number of farmers relative to the area cultivated, and the existence, as noted, of a sizable number of growers with large holdings, would suggest that it was no longer easy for a young man to become a coffee grower. The 1973 survey suggested that only 3 per cent of the growers in the Eastern Highlands were under 25 years of age, and in 1976 the young proportion of farmers in the highlands region was far below that for the general population. [227]

The weak position of the mass of farmers is also suggested by the fact that, although some 80 per cent of all adults in the Eastern Highlands were engaged in coffee production to some extent in 1976, the average smallholder income per head from all export crops in the province in the same year was only about K30. [228] The per capita annual income of Gorokans in the early 1960s was some $2.50; in the mid 1960s about $10; and was some $20 at the end of that decade. [229] These income characteristics are important but are only loosely suggestive of the comparatively weak positions of the poor and middle peasants.

In less than twenty years, the colonial state had assisted the formation of a powerful rich peasant class with roots deep in ancillary pre-capitalist and 'subsistence' modes of production. The strength of the rich peasant class was its ability to present itself to those it dispossessed and manipulated as, at one and the same time, both 'traditional' and 'modern'. For those to whome some of the wealthy classes contemptuously refer as 'bush <u>Kanakas</u>' the formation of the rich peasantry represented a new road, and a chance for improved material and cultural conditions. The new class carries with it the culture of previous generations, of which, through its co-option of the big man institution, it is the guardian and interpreter.

For the colonial state the new class proved the 'correctness' of their administrative and, particularly, agricultural policies. Just as had been hoped, an elite group of 'progressive' farmers was 'showing the way' to the bulk of 'backward' villagers.

For the rich peasants themselves, the years preceding the seventies were years of promethean labour, and of immense political and personal risk. Traditional advantage plus an ability to seize the time enabled members of the old satrapy to take the opportunity to accumulate wealth and power created by an active and interventionist colonial state, and through the 1970s, to extend and entrench their position and privilege.

Notes

[1] G.F. Vicedom and H. Tischner, Die Mbowamb, Hamburg, Cram, de Gruyter and Co., 1943-8.
[2] Abraham L. Gitlow, Economics of the Mount Hagen Tribes, New Guinea, Seattle and London, University of Washington Press, 1947, p. 83.
[3] Ibid.
[4] Ibid., pp. 35 and 37.
[5] Ibid., p. 81.
[6] E.A. Young, 'Population Mobility in Agarabi/Gadsup, Eastern Highlands Province', in R. May (ed.) Change and Movement: Readings in Internal Migration in Papua New Guinea, Canberra, ANU, p. 179.
[7] Ibid.
[8] I.L. Idriess, Gold Dust and Ashes. Sydney: Angus and Robertson, 1934, pp. 103-4. J. Sinclair, Kiap Australia's Patrol Officers in Papua New Guinea, Sydney, Pacific Publications, 1980, pp. 43 and 45.
[9] Idriess, ibid., p. 137.
[10] Sinclair, op. cit., p. 43.
[11] Ibid., p. 75.
[12] I. Downs, The Australian Trusteeship in Papua New Guinea 1945-1975, Canberra, AGPS, 1980, p. 183.
[13] Sinclair, op. cit., pp. 254-5.
[14] Ibid., p. 71.
[15] Ibid., p. 45.
[16] F. West, 'An Australian Moving Frontier in New Guinea', in N. Gunson (ed.), The Changing Pacific: Essays in Honour of H.E. Maude, Melbourne, Oxford University Press, 1978, p. 223. S.W. Reed, The Making of Modern New Guinea, Philadelphia, The American Philosophical Society, 1943, p. 237.
[17] R.J. Lacey, 'Missions' in P. Ryan (ed.), Encyclopaedia of Papua New Guinea, Clayton, Melbourne University Press, 1973, pp. 778-9.
[18] Reed, op. cit., p. 41.
[19] Ibid., p. 203.
[20] Sinclair, op. cit., p. 73.
[21] R. Radford, 'Burning the Spears: A 'Peace Movement' in the Eastern Highlands of New Guinea 1936-7', Journal of Pacific History, 12, 1, 1972, p. 46.
[22] Gitlow, op. cit., p. 45.
[23] Ibid., p. 47.
[24] Ibid., and Sinclair, op. cit., p. 173.
[25] R.F. Salisbury, From Stone to Steel, Clayton, Melbourne University Press, 1962a, p. 123.
[26] J.D. Legge, Australian Colonial Policy: A Survey of Native

Administration and European Development in Papua, Sydney, Angus and Robertson, 1956, pp. 202-3.

[27] F.J. West, 'Indigenous Labour in Papua New Guinea', *International Labour Review*, 77, 2, 1958, p. 99.

[28] *Ibid*., p. 96.

[29] Downs, *op. cit*., pp. 38 and 39.

[30] West, *op. cit*., p. 96.
'Where insufficient voluntary labour was forthcoming compulsion was used, and that became the rule rather than the exception'. Some New Guinea villages were 'almost denuded' of labour. Legge, *op.cit*., pp. 185 and 186.

[31] Downs, *op. cit*., p. 39.
The following song is translated from Motu, and was sung by indentured workers during the War:
Song that is sung in remembrance of homes left behind
We have left our homes and beaches
To labour for the war in difficult places,
In far flung places. In these hard times
We wander aimlessly away from home.
Places that we have never seen before
We now have seen by toil and sweat.
Who has caused this dreaded parting
To be out on scattered places?
H.I. Hogbin, *Transformation Scene: The Changing Culture of a New Guinea Village*, London, Routledge and Kegan Paul, 1951, pp. 13-14.

[32] West, *op. cit*., p. 96. See also Good and Peter Fitzpatrick 'The formation of the working class', in Amarshi, Good and Mortimer, *Development and Dependency*, Melbourne, Oxford University Press, 1979, pp. 129-130.

[33] *Ibid*, p. 97.

[34] Downs, *op. cit*., p. 46.

[35] West, *op. cit*., p. 98.

[36] Downs, *op. cit*., p. 47.

[37] Sinclair, *op. cit*., p. 173.

[38] West, *op. cit*., p. 100.

[39] West, *op. cit*., p. 102. Sinclair, *ibid*.

[40] Downs, *op. cit*., p. 48. Ashton, *op. cit*., p. 191.

[41] C. Ashton, 'The Leahy Family' in J. Griffin (ed.), *Papua New Guinea Portraits*, Canberra, ANU, 1978, p. 148.

[42] West, *op. cit*., p. 104.

[43] *Ibid*., p. 105.

[44] Hogbin, *op. cit*., p. 187.

[45] *Ibid*., pp. 193 and 192.

[46] Salisbury, 1962a, *op. cit*., p. 126.

[47] West, *op. cit*., p. 109.

[48] *Ibid*., p. 107.

[49] I. Downs, '*Kiap*, Planter and Politician: a Self Portrait' in J. Griffin (ed.), *Papua New Guinea Portraits*, Canberra, ANU, 1978, p. 228.

[50] West, *op. cit*., p. 107.

[51] Salisbury, 1962a, *op. cit*., p. 132 note 10.

[52] Hogbin, *op. cit*., p. 195.

[53] Salisbury, 1962a, *op. cit*., pp. 128 and 131.

[54] West, *op. cit*., p. 107.

[55] 'For a long period we have quite wisely tried to maintain villages intact from outside disturbances. We are still discouraging

disturbance or the depopulation of villages ...' P. Hasluck, 'Present Tasks and Policies', in J. Wilkes (ed.), New Guinea and Australia, Australian Institute of Political Science, Angus and Robertson, 1958, p. 109.

[56] R.F. Salisbury and M.E. Salisbury, 'The Rural-Oriented strategy of Urban Adaptation: Migrants in Port Moresby' in R. May (ed.), Change and Movement: Readings in Internal Migration in Papua New Guinea, Canberra, ANU, 1977, p. 221.

[57] Ibid., p. 222.

[58] Ibid., p. 225.

[59] West, op. cit., p. 106.

[60] Salisbury and Salisbury, op. cit., p. 223. E. Brown, 'The Kainantu Villager', Journal of the Papua New Guinea Society, 3, 1, 1969, p. 62.

[61] Downs, op. cit., p. 165.

[62] H.C. Brookfield, 'Native Employment within the New Guinea Highlands', Journal of the Polynesian Society, 70, 1961, p. 311.

[63] Hogbin, op. cit., p. 180.

[64] H.I. Hogbin, 'Local Government for New Guinea', Oceania, 17, 1, 1946, p. 43.

[65] Ibid., p. 62.

[66] Langness, 1963, op. cit., p. 154.

[67] Sinclair, 1980, p. 39.

[68] Downs, 1980, p. 138.

[69] A.Strathern, 'Kiap, Councillor and Big Man: Role Contrasts in Mt. Hagen', in M.W.Ward (ed.), The Politics of Melanesia, Papers Delivered at the Fourth Waigani Seminar, Port Moresby and Canberra, University of PNG and ANU, pp. 554 and 558.

[70] Ibid., p. 555.

[71] Ibid., p. 556.

[72] Ibid., p. 559.

[73] Ibid., p. 564.

[74] M. Reay, 'Present-Day Politics in the New Guinea Highlands', New Guinea the Central Highlands, American Anthropologist, 66, 4, 1964, p. 250-1.

[75] Legge, op. cit., p. 219-220.

[76] L.P. Mair and I. Grosart, 'Local Government', in P. Ryan (ed.), Encyclopedia of Papua New Guinea, Clayton, Melbourne University Press, 1972, p. 657.

[77] Downs, 1980, op. cit., p. 102.

[78] Reay, 1964, op. cit., p. 251.

[79] Downs, 1980, op. cit., pp. 102 and 201.

[80] Hasluck, op. cit., p. 112.

[81] Mair and Grosart, op. cit., p. 658.

[82] Hasluck, op. cit., p. 106.

[83] P. Hasluck, 'The Economic Development of Papua and New Guinea', Australian Outlook, 16, 1, 1962, p. 24.

[84] Downs, 1980, op. cit., p. 106.

[85] M. Reay, 'Roads and Bridges Between Three Levels of Politics', in M. Ward (ed.), The Policits of Melanesia. Papers Delivered at the Fourth Waigani Seminar, Port Moresby and Canberra, University of PNG and ANU, 1973, p. 544.

[86] J.T. Gunther, 'The People', in J. Wilkes (ed.), New Guinea and Australia, Australian Institute of Political Science, Angus and Robertson, 1958, p. 64.

[87] Sinclair, 1980, op. cit., p. 256.

[88] Mair and Grosart, op. cit; Downs, 1980, p. 340.

[89] Ibid., p. 662.
[90] Ibid.
[91] Ibid.
[92] Reed, op. cit.
[93] R. Radford, 'Aiyura Research Station', History of Agriculture Discussion Paper Number 35, Department of Primary Industry and History Department, UPNG, 1979, pp. 6-7.
[94] Ibid., p. 1.
[95] Ibid., p. 2.
[96] Ibid., p. 4.
[97] Ibid.
[98] P.M. Munster, 'The Ground of the Ancestors. Part One of a History of Goroka', M.A. Qualifying Sub-Thesis, History Department, UPNG, 1973, p. 41.
[99] Bjorn Beckman, Organising the Farmers, Uppsala, Scandanavian Institute of African Studies, 1976, p. 37.
[100] Finney, op. cit., p. 42.
[101] Ashton, op. cit.
[102] Gitlow, op. cit., p. 74.
[103] Finney, op. cit., Salisbury, 1962a, op. cit., p. 125.
[104] Downs, 1980, op. cit., p. 174.
[105] Downs, 1978, op. cit., p. 240.
[106] Sinclair, op. cit., p. 254.
[107] Ibid., p. 41.
[108] Downs, 1978, op. cit., p. 244.
[109] P. Hasluck, 'Lands Policy in Papua and New Guinea', Statement by Minister for Territories, South Pacific, 7, 12, 1954.
[110] K.E. Read, 'Land in the Central Highlands', South Pacific, 6, 7, 1952.
[111] M. Donaldson and K. Good, 'The Eastern Highlands, Coffee and Class' in D. Denoon and C. Snowden (eds.), A Time to Plant and a Time to Uproot: A History of Agriculture in Papua New Guinea, Port Moresby, Institute of Papua New Guinea Studies, 1980, p. 147.
[112] Ibid.
[113] Finney, op. cit., p. 685, note 5.
[114] Donaldson and Good, 1980, op. cit.
[115] Ashton, op. cit., p. 192.
[116] Ibid, p. 192-3.
[117] Ibid., p. 193; Downs, op. cit., p. 249; Sinclair, op. cit., p. 255.
[118] Radford, 1979, op. cit., p.3, note 14.
[119] Donaldson and Good, 1980, op. cit., p. 147-8.
[120] Ibid., p. 148.
[121] J. Cartledge, A History of the Coffee Industry in Papua New Guinea, 1978.
[122] N. Bluett, 'Report at the District Staff Conference', DASF Eastern Highlands, Mimeo, November 1969, DPI Archives, Goroka.
[123] Finney, op. cit., p. 54.
[124] R. Southern, 'Road Transport in the New Guinea Highlands', Department of Geography Occasional Paper Number 6, UPNG, 1973, p. 28.
[125] Finney, op. cit., pp. 42-47, estimated that there was a ten-fold increase in the area alienated.
[126] Ibid., pp. 48-56.
[127] Ibid., p. 53.
[128] Ibid., p. 55.

[129] Post Courier, 31 May, 1977, p. 14.
[130] Finney, op. cit., p. 54.
[131] Ibid., pp. 168-9.
[132] Quoted in Good, 1979, 'The Formation of the Peasantry', in Amarshi, Good and Mortimer, op. cit., p. 119.
[133] Quoted in Finney, op. cit., p. 54.
[134] Munster, op. cit., pp. 15 and 40.
[135] P.G. Sack, 'Land Law and Policy in German New Guinea', Australian External Territories, 1971, 11, 1, p. 17.
[136] Quoted in Munster, op. cit., p. 20.
[137] Letter from W.L. Conroy, a/chief, Division of Agricultural Extension, DASF, Port Moresby to district agricultural officer, Goroka, 6 September, 1965, file 1-1-A, DPI Archives, Goroka. Barrie, however, in 1956, referred to a 'minimum of half an acre or three hundred trees' as 'the minimum economic unit'.
See J.W. Barrie, 'Coffee in the Highlands', Agricultural Extension Conference, Goroka, November 1956, pp. 1 and 15, DPI Archives, Goroka.
[138] Barrie, ibid., p. 2.
[139] Finney, op. cit., p. 93.
[140] Ibid.
[141] Finney, op. cit., pp. 91-92.
[142] Ibid.
[143] Ibid.
[144] H. Brookfield, 'The Money that Grows on Trees: the Consequences of an Innovation within a Man-Environment System', Australian Geographical Studies, 6, 2, 1968, p. 105.
[145] Barrie, op. cit., p. 16.
[146] R.F. McKillop, 'A History of Agricultural Extension in Papua New Guinea', in G. Dick and McKillop, A Brief History of Agricultural Extension and Education in Papua New Guinea, Extension Bulletin Number 10, DPI, 1976, p. 24.
[147] G. Harris, 'The Role of the Agricultural Extension Officer: a Study in Eight Highlands Villages', in Harris and von Fleckenstein Agricultural Extension in the Village, Extension Bulletin No. 9, DPI, 1975, p. 16.
See also S. Epstein, 'Innovation of Cash Crops in New Guinea Subsistence Economies', Economic Record, 46, 1970.
[148] R.N. Amos, Assistant District Agricultural Officer, 'Summary of Primary Industries in the Eastern Highlands District', no date. File 1-2-A, DPI Archives, Goroka.
[149] A. Amarshi, 'The Economy: the Development of Peripheral Capitalism', in Amarshi et. al., op. cit., p. 37.
[150] McKillop, quoted in Good, 'The Formation of the Peasantry', p. 112.
[151] Quoted in R.J. Cheetham, 'The Development of Indigenous Agriculture, Land Settlement and Rural Credit Facilities in Papua New Guinea', PNG Agricultural Journal, 15, 3-4, 1962-63, p. 69.
[152] Letter from Henderson, acting Director of Agriculture, to the District Commissioner in Goroka, 10 April, 1956, DPI Archives, Goroka.
[153] McKillop, 1976, op. cit., p. 25.
[154] Finney, op. cit., p. 66.
[155] Maurice Godelier, 'The integration of ideology, methodology and practice in field research: case study - Baruya', seminar in the Anthropoloty and Sociology Department, University of PNG, 16 October, 1978.

[156] See Donaldson and Good, 'Class and politics in the Eastern Highlands of Papua New Guinea', History of Agriculture Discussion Paper Number 9, DPI and UPNG, 1978.

[157] J.L. Anderson, 'The development of a cattle industry in the territory of Papua and New Guinea', Papua New Guinea Agricultural Journal, 14, 4, 1962, p. 139.

[158] DASF Annual Report 1952, and ibid.

[159] R.F. McKillop, Report on Native Cattle Projects in the Goroka and Chimbu sub-districts, June 1963, File 14-3-3 (5), DPI Archives, Goroka.

[160] Ibid.

[161] Letter from Henderson, Director DASF, to District Commissioner, Eastern Highlands, dated 12 May, 1960, File 14-3-3(5), DPI Archives, Goroka.

[162] McKillop 1976, op. cit., p. 30.

[163] Ibid.

[164] Eastern Highlands District Advisory Council, Minutes and Appendices, Meeting at Goroka, 21 May 1965, DPI Archives, Goroka.

[165] Quoted in Donaldson and Good, 1980, op. cit., p. 159.

[166] Ibid.

[167] Ibid.

[168] J. Lloyd, 'Land tenure in Papua and New Guinea', Australian External Territories, 11, 1, 1971, p. 4.

[169] Report on the Commission of Inquiry into Land Matters, Port Moresby, October 1973, p. 25.

[170] Jim Fingleton, 'Report on blocks tenure converted under section 7', mimeo.

[171] Ibid.

[172] J.R. Gusfield, 'Tradition and modernity: misplaced polarities in the study of social change', American Journal of Sociology, 72, 4, 1967, p. 358.

[173] R. Crocombe, 'Progress means making changes', Post Courier, 18 July, 1978.

[174] Fingleton, op. cit.

[175] D.S. Grove, 'Land use and Land Tenure - the New :egislation', in M. Ward (ed.), Change and Development in Rural Melanesia, 1972, p. 70.

[176] McKillop, 'Native Owned Cattle in the Eastern Highlands - a View to Date', no date, DPI Archives, Goroka.

[177] Quoted in Donaldson and Good, 1980, op. cit., p. 160.

[178] McKillop, 'EHD Cattle Projects Annual Report 1965/66', DPI Archives, Goroka.

[179] P. Dowling, 'Land Policy and Economic Development', Australian External Territories, 11, 1, 1971, p. 39.

[180] McKillop, 1976, op. cit., p. 34.

[181] The following table illustrates similar rates of growth over a longer period of time and with a smaller group of people. The Numuyargobo Confederacy comprised almost 2,000 people from three 'tribes' in the Bena census division. The people of the area participated in one of the pilot cattle projects.

Table 4.10
The growth of the cattle industry in Numuyargobo 1960-70

	1960	1962	1966	1967	1970
Number of Projects	2	3	6	6	18
Number of Cattle	9	26	92	121	383
Area fenced (acres)	10	15	146	176	3569
Average number acres per project	5	5	24	29	198

Source: R. McKillop, An Agricultural Development Strategy, the Numuyargobo Cattle Project in Papua New Guinea, Extension Bulletin No. 2, DPI, 1973, p. 3.

[182] Cheetham, op. cit., p. 74.
[183] Ibid., p. 75.
[184] It was not well endowed with funds and it was expressly intended as a lender of last resort. Four Australian banks operated but lent chiefly to foreign plantations and businesses. M.J.P., 'Financial Institutions Other than Banks', 1973; and K.R. Woodward, Agriculture in the Economy: Rural Credit and Foreign Aid, DPI, Port Moresby, no date.
[185] A.P.J. Newman, 'Native Loans Fund; Notes for Guidance of Officers Assisting with Loan Applications', no date, DPI Archives, Goroka.
[186] Letter from P.R. Harvey, Division of Animal Husbandry, Goroka, to district agricultural officer, 22 August 1963; and B.A. Boniwell, 'District Policy', 30 August, 1963. File 1-1-C and 1-2-A, DPI Archives, Goroka.
[187] R.V. Kumaina, 'Savings and Loans Societies in Papua New Guinea'. Paper presented at the Third Waigani Seminar, Port Moresby 1964, p. 3 and Appendix 1.
[188] Ibid., p. 4.
[189] Ibid.
[190] Eastern Highlands District Advisory Council, op. cit.
[191] K.G. Crellin, 'The Use of Credit Facilities in Rural Areas' in M. Ward (ed.), Change and Development in Rural Melanesia, 1972, pp. 513-514.
[192] Development Bank, Administration Agency Circular, Numbers 4, 6, 1968, and 15, Port Moresby.
[193] Ibid, Numbers 5, 6, January 1968.
[194] DASF Goroka district Annual Report, 1970-71, and Good op. cit., p. 113. Cattle, nationally, was the biggest recipient of Bank loans; to the end of June 1976, over one-third of the Bank's lending, or more than K20 million, went to cattle (next highest sector was oil-palm with K4 million); Woodward op. cit., p. 7. And cattle growing necessarily meant, in DASF's view, relatively large-scale land ownership - projects of less than 50 acres were deemed uneconomic because of their low potential stocking rate; Letter from P.T. Bowater, rural development officer, Korofeigu to PNG Development Bank, 21 December 1970, File 9-1-D, DPI Archives, Goroka.
[195] Development Bank, op. cit., Circular number 6, 28 March 1968.
[196] Letter to R. McKillop from the manager, Operations Department of PNG Development Bank, 30 June 1970. DPI Archives, Goroka.
[197] The Bank's relative flexibility of approach to land matters was accompanied by an underlying firm preference for freehold ownership. As a Development Bank document declared:
The best possible assurance of continued land usage is, of

The best possible assurance of continued land usage is, of course, the possession of freehold (converted tenure), or leasehold title ... whilst we fully support any spontaneous desire by a loan applicant for tenure conversion ... we try not to hold up approval and implementation of the loan because of the time necessarily involved in conversion proceedings ... To cover the situation the Bank has adopted a procedure of obtaining evidence of clan agreement to land usage ... it is quite sufficient for our purposes in most cases, provided responsible Administration officers in the area(preferably including D.O. Lands) are satisfied that it represents the genuine attitude of the clan concerned.

PNG Development Bank, Land Usage Agreement, no date.

[198] Development Bank, op. cit., Circular Number 5, January, 1968.

[199] Rural Loans, PNG Development Bank, DPI File 9-1-D DPI Archives, Goroka; and 1977 National Election, Notes on Candidates.

[200] DASF District Annual Report, 1970-71, and Rural Loans File 9-1-D, DPI Archives, Goroka; and inteview with Masuwe by Jim Fingleton, op. cit.

[201] Development Bank, op. cit., Circular Number 20, 23 April 1971.

[202] McKillop, op. cit., 1976, p. 34.

[203] Ibid, p. 44.

[204] Crellin, op. cit., pp. 511, 509.

[205] W. Black, 'New Guinea Cattlemen', Australian External Territories. 10, 4, 1970, p. 19.

[206] Kindino obtained his large loan through the livestock branch (which was particularly inclined to encourage patron-client relations) in Lae, so that when he moved to the Eastern Highlands he moved outside the normal district orientated extension sevice. He became President of the Kainantu Cattle Owners' Association, but as an outsider, and thus in not a big man, he became the focus ofhositility. His tractor was rolled and his cattle speared and he died in about 1978 because, it was said, someone put battery acid in his beer (personal communication, McKillop, 1978).

[207] F. von Fleckenstein, 'Ketarovo: Case Study of a Cattle Project' in Harris and von Fleckenstein, 'Agricultural Extension in the Village', Extension Bulletin Number 9, 1975, p. 60.

[208] Finney, op. cit., pp. 92-95 and 101.

[209] McKillop quoted in Good, 'Formation of the Peasantry', p. 113.

[210] Post Courier, 11 September 1974.

[211] Eg., Teodor Shanin, 'Peasantry as a Political Factor', in Shanin (ed.), Peasants and Peasant Societies, Harmondsworth, Penguin, 1971.

[212] Finney, op. cit., p. 99.

[213] D. Anderson, An Economic Survey of Smallholder Coffee Producers - 1976, Port Moresby, 1977, p. 25.

[214] Interview by Aita Ivarato with Kangeto Yavise at Henganofi, 28 June 1978. From late 1975 until early 1977 the legal rural minimum wage was about K463 per annum, being lower in primary industry and higher in other rural activities in the provincial towns. National Minimum Wages Board, 'Reasons for decision', Port Moresby, June 1976, pp. 60-63.

[215] Finney, op. cit., pp. 96-97, and 85 (where Finney says that Akunai had 18,000 coffee trees); and application for Development Bank Loan, from McKillop, 3 January 1968. It was noted that Akunai had obtained a loan from the Native Loans Board in 1960 'to pay his labourers', File 9-1-D, DPI Archives, Goroka; Good, op. cit.;

and interview with John Akunai by Jim Fingleton, op. cit.

[216] D. Anderson, op. cit., pp. 25, 16.

[217] John Iliffe, A Modern History of Tanganyika, Cambridge, Cambridge University Press, 1979, chapter 9.

[218] Finney, op. cit., p. 85.

[219] D. Anderson found that the average holding in the Western Highlands was significantly larger than in the Eastern Highlands. An Economic Survey of Smallholder Coffee Producers, p. 10.

[220] Ibid., p. 16.

[221] Ibid., pp. 18-19.

[222] Figures for 1973 from L.D. Wilson and G.B.A. Evans, Sample Survey of Smallholder Coffee Producers, DASF, 1975, p. 11. 1976 figures from DPI Goroka (personal interview with J.J. Nitsche, 4 July 1977).

[223] Eastern Highlands population figures from Population Characteristics Bulletin No. 10 - Eastern Highlands District, and from 'Population projections, by province', Bureau of Statistics, Port Moresby and Waigani. The projections are believed to err on the high side and the estimates of the size of the rich peasantry in the province in 1976 are therefore both approximate and generous.

[224] Finney, op. cit., p. 85.

[225] Wilson, op. cit., p. 14, and D. Anderson, 'Summary of results ...' September 1975, Mimeo.

[226] Personal interview with J.J. Nitsche, DPI, Goroka, 4 July 1977.

[227] Wilson and Evans, op. cit., p. 12, and D. Anderson, An Economic Survey ..., p. 9. The two reports differ a little on their estimates.

[228] The National Public Expenditure Plan 1978-1981, Waigani, National Planning Office, February 1978, Exhibit 3.5, p. 54.

[229] World Bank, op. cit., p. 11, and Diana Howlett, 'Terminal Development: from Tribalism to Peasantry', in Harold Brookfield (ed.), The Pacific in Transition: Geographical Perspectives on Adaptation and Change, Canberra, ANU Press, 1973, p. 266.

5 Rural capitalism and the successor state

With national independence in 1975 the rich peasants were no longer just clients of the state, but in certain key respects and areas its influential constituents as well. The future of the rich peasantry as a class as it moved through the 1970s rested upon a number of important factors, some open to its class action, some beyond its control. These included the vagaries of world commodity markets; the degree of control it maintained over the provincial and national governments; the ability of members of the class to voice, promote and defend the interests of the class as a whole; the extent to which it was able to control and direct the very process of class formation itself in retarding the cohesion of developing oppositional classes; and its ability to prevent the emergence within itself, or at least to contain, factional and other differences.

The wealth of leading rich peasants in the late 1970s in the Eastern Highlands was visible in graphic detail. Hari Gotoha, for example, was last on Finney's list of business leaders near the end of the 1960s, with assets of 1,400 coffee trees, one trade store, two trucks and ten cattle, [1] but he subsequently acquired ownership of a small Gorokan taxi company, and a tyre and battery service and became the largest single national shareholder in Pacific Helicopters based in Goroka (holding in 1977 shares worth K10,000). [2]

More importantly perhaps, Gotoha set about the formation of the Gouna Development Corporation, towards the mid 1970s. A few years later, the Gouna Corporation owned some 260 acres of coffee (its largest plantation, Lahamenegu, of 200 acres, was formerly held by Jim Leahy), a poultry farm with about 9,000 birds, a modern piggery and, as well, a Goroka book and sports store and a shopping arcade (held in association with Collins and Leahy with Gouna owning 55 per cent of the property). Total fixed assets were valued at K700,000. [3] Gotoha's personal property and that of Gouna Corporation had become closely intertwined.

Ownership of the new Goroka soft drink company ('Laga') was 30 per cent Gotoha personally and 15 percent Gouna, and Goroka Tyre and Battery was both a Gouna firm and Gotoha's property and a separate shareholder in Pacific Helicopters. Gotoha was himself secretary of the corporation, and he and his close associates possessed a controlling influence over its affairs. Gotoha, Auwo Ketauwa (chairman), and Isembo Olekavea (vice-chairman), along with the paid expatriate manager, constituted the executive committee of the board of directors, and made the effective decisions. Hari Gotoha was thus a wealthy man whose property holdings were characterised by diversity and by a strong personal interest and preponderant influence in the large and successful Gouna Corporation.

Auwo Ketauwo and Gotoha came together in the early 1970s, and shared the aim of purchasing foreign owned businesses then up for sale and of emulating the New Guinea Development Corporation founded by the Mataungan Association in Rabaul, and headed by John Kaputin (a cabinet minister until 1982 in each government since independence). Ketauwo became president of the Federation of Savings and Loans Societies and a director of Summerscales and Lambert, Goroka's leading building company, and he acquired a truck, two trade stores in Goroka, and a shareholding in Pacific Helicopters. [4] Later he had registered 103 hectares of land in his name and established a large coffee farm, and also gained an interest in some 700 hectares of grazing land and 150 cattle in the Madang area. [5] Auwo Ketauwo was therefore an experienced businessman, a strong influence in Gouna, and a newly established but large-scale coffee and cattle farmer.

Nor was Isembo Olekavea a negligible figure with, in 1978, five hectares of coffee, a further 97 hectares of registered land, and an income of some K7,000 in the previous year. He was also one of ten members of the PNG Coffee Industry Board in Goroka, and the only member from the Eastern Highlands. [6] Paul Ine, another Gouna director, in 1978, not only owned some 4.5 hectares of coffee and controlled some 22.5 hectares of cattle land, but also possessed three trade stores and two trucks. He subsequently established Goroka's first clothing factory at a cost of K170,000, employing initially 30 young women and with plans for another 70. Paul Ine too was very well placed organisationally in 1978, being simultaneously president of the Goroka chamber of commerce, minister for commerce in the interim provincial government (IPG), chairman of directors of Nokondi, the provincial government's business arm, and an influential director of the Gouna corporation. [7] Gotoha, Ketauwo, and Ine were therefore similar in their strong and diversified commercial and farming interests, and together with Olekavea they commanded positions in important commercial and political organisations. [8]

Akepa Miakwe and Sailas Atopare were also to be included among the most successful of the rich peasantry in the Eastern Highlands in the late 1970s. Miakwe near 1978 had some ten hectares of coffee with about 7,000 trees, two coffee pulpers, a commercial vegetable garden, trucks, and a trade store; and he was a shareholder and director of Namasu Coffee, a coffee exporting company, a director and virtual founder of the Bena (later called the Eastern Highlands) Development Corporation, and a member of the House of Asembly since 1972. He claimed an annual income from all sources in 1976 of some K30,000. [9] Atopare was, in his twenties, the youngest man referred to here, but he too owned some 25 hectares of coffee, a commercial vegetable garden, a trade store, a bus and two cars, and he was a shareholder in nine firms, and the owner of a flat in Port Moresby. He was additionally a director of the

Asaro Watabung Rural Development Corporation and, since 1977, like Miakwe a member of the House of Asembly. He claimed a total income in 1977 well in excess of K10,000. [10] These men too are thus active in both farming and commerce, and each was well placed within a development corporation and was a member of the national parliament to boot. All six men - Gotoha, Ketauwo, Olekavea, Ine, Miakwe and Atopare - were of course employers of wage labour in both their private and their corporate capacities.

The comparative wealth of such men and in particular their diversified economic activity suggests that a new class of rural capitalists was emerging in the Eastern Highlands through the 1970s. In simple property terms they owned considerably more than 1,000 trees and a coffee pulper designating the base line of the rich peasantry, and all employed permanent wage labourers. The differentiation between the rich peasantry, representing perhaps some six per cent of highlands' coffee growers, and the far smaller class of rural capitalists is in terms of the extent of its property holdings, control over coffee processing and buying and the economic diversification associated with the latter. Considerable overlaps would exist between the two social entities, as between the rich and middle peasantries. Social relations within a developing rural capitalism are in the process of reformation where, at any one time, 'unformed' classes may exist. [11] Classes exist in varying states of formation and action, where what classes (or their leading elements) do, shape the composition and size, the formation, of the classes themselves. The rich peasantry is only the most visible and palpable social entity in the highlands at present, but a new class of rural capitalists is in the process of formation and consolidation, and the existence of the rich peasants requires the existence of middle and poor peasants. The rich peasantry and rural capitalists might also be referred to simply as the rich rural classes, on the one hand, as against the mass of the peasantry, on the other. [12]

The big holdings of the rich rural classes in the Eastern Highlands also point to the irrelevance of continuing to consider all national coffee growers ('smallholders') together in simple contrast to the supposedly foreign plantation owner ('largeholder'). The range of indigenously owned coffee holdings both private and corporate is now large indeed, and the notion of the smallholder, if applied to this whole range, obscures the growing wealth and dominance of the rich classes. This is particularly the case when, as we shall see in the following chapter, the rural capitalists proceeded to take over the 43 'largeholdings' in the province. These, averaging 55 hectares and ranging from 10-400 hectares in size, accounted for 40 per cent of coffee production in the Eastern Highlands in 1979. [13] In that year, 'largeholdings' which were formerly defined by DPI in terms of their ownership by non-Papua New Guineans, were redefined to mean holdings over 10 hectares. [14]

While the rich rural classes increased their wealth, and the holdings of the mass of coffee growers in the Eastern Highlands declined, the inequalities between the two obviously widened. The province's relative health status index in 1976 was low, as it was for all other highlands' provinces, and its educational status in the same year was second from the bottom of the national list, with only Southern Highlands Province in a slightly worse positon. [15] Shortages of land and of locally available labour were being experienced in the Goroka area as early as the late 1960s, when Finney reported that 'many a young man has difficulty in getting even a little land for a modest coffee garden', [16] and the situation has undoubtedly worsened since then over a far wider area of the province. It is often the case in colonial and post-colonial situations that, where land is given over to the production of export crops, falling food resources for local consumption are experienced. Problems of food, health, land and education are, however, not experienced on a socially equitable basis but on a highly differentiated one, where the rich rural classes utilise their socio-political privileges, as it were, to pass the problems on to the poorer classes. The holdings of the rich classes increase as those of the small coffee grower decline, and much the same process seemed to be at work in the increasing incidence of beer consumption and deteriorating nutritional standards widely experienced in the Eastern Highlands in the late 1970s. [17]

Out of and beyond the co-operatives

Classes are properly defined and understood not only structurally but also in the actual experience of class action and struggle, activities which express the common interests of members of the particular class and define them against the interests of other classes in the social formation. The rising rich peasantry of the Eastern Highlands has defined itself as a class through its action over recent years.

The exploitation of clan members' labour and land, as already noted, was an important technique for initiating the career of ambitious and influential farmers, and the tapping of villagers' savings was a similar device resorted to before the establishment of state bank lending, and subsequently maintained in some significant cases, the development corporations for one, as a supplement to the banks. The manipulation of the funds and equipment of co-operative societies (generally referred to as Rural Progress Societies), however, clearly expressed the interests of rising rich peasantry distinct from other community members, and therefore deserves further attention.

Co-operatives also existed as pressure groups for state aid for rising farmers, but to a large extent aid from the Department of Agriculture, Stock and Fisheries (DASF) as noted was freely given, to some. With the Yagaria Progress Society, DASF in January 1965 made a tractor and plough available to prepare land for peanut planting in areas of between two and twelve acres, specifically as family lots for the directors of the society. Two months later DASF followed this with the loan of a grader free of charge. The power of the kiap was turned to, and he agreed that the directors would have the right to fine the owner of any pig damaging the peanut gardens. [18] The leading men of the co-operative were conscious of their position vis a vis both the ordinary villager and the ordinary society member. Honoraria were paid to directors in early 1965, and later specially designed badges were ordered at a cost of $110 for 25, which were distributed at a ceremony to which dignitaries such as local members of the House of Assembly were invited. [19] Particular men did particularly well. James Yanepa, previously a DASF farmer-trainee, became a full time employee of the society in 1965, at a salary of 4 pounds per fortnight. He was made the society clerk and secretary and received a new house in which he kept the sometimes large amounts of cash necessary for the society's purchases of peanuts and coffee and the proceeds resulting from the sales. Later he successfully requested permission to attend an agricultural course at Korofeigu while he contined to be paid by the society, but returned unexpectedly to contest a by-election for the House of Assembly. [20]

While James Yanepa was moving ahead individualistically, the leading members showed their group consciousness by restricting members' voting rights. At a meeting in July 1966 when there were 245 shares of $10 each and 317 part shares (for a total share capital of $3,799), the society's rules were changed to permit only those with ten shares or more to vote. [21] Other action by leading members can be deduced from the malaise that became noticeable in the society, and from the reported 'general underground movement against the Secretary', Yanepa, in early 1971, when a 'considerable number' of members were refusing to sell coffee to him. The secretary was then known to have acquired among other things, over less than six years, a trade store, a new trucking business and was planning a pig farm. [22] A report to the Assistant Registrar of Co-operative Societies, 22 October 1971 said, 'the general attitude among Yagaria Progress Society members is that the original share capital was squandered by those who were supposedly assisting and advising the society'. [23] The loss of members' money and the collapse of the co-operative did not of course adversely affect James Yanepa, who went on to become the first Premier of the Eastern Highlands Provincial Government.

Events within the Yagaria Progress Society were not atypical of co-operatives in the Eastern Highlands. A number of peculiarities occurred in the use by rising farmers of machinery that was supposedly owned by the Numuyargobo Livestock Co-operative, and in the purchase of a tractor in 1967, certain villagers were reportedly 'taken for a ride' by the secretary of the Numuyargobo Society, Nou Matea, and another man. [24] Another official report noted that the main aim of potential share-holders was to use the Numuyargobo Society as a pressure group to obtain a large supply of cattle for themselves, and in July 1970 it was recorded that the co-operative possessed more machinery than share capital. In an area with a population of some 4,000 its membership was only 254. [25] The members of the Numuyargobo co-operative were exhorted directly by an unnamed DASF official in 1969 to make themselves 'prosperous businessmen' as quickly as possible or remain 'rubbish men'. [26] Some at least appear to have taken the advice to heart.

Through the holding of directorships in important rural societies, the rich rural classes are able to widen their economic activities in perfectly legitimate ways. Both Hari Gotoha and Auwo Ketauwo were directors of the Eastern Highlands Savings and Loans Society which, in 1977, possessed total savings of K1.7 million. In the same year, Gotoha as chairman reported that the society's assets had more than doubled on the previous year and stood at over K2 million. Loans were made to 20 per cent of the 6,385 members and averaged K650 per loan, although one member was able to borrow K100,000 to purchase a business. [27] The furthering of personal interest which could occur through the control of such societies was suggested by Ketauwo who is additionally president of the PNG Federation of Savings and Loans Societies (assets of K10.4 million in 1976), and who saw the rules limiting borrowing from the society by its directors as 'bad for development'. [28] However, as of early 1978, twenty six directors of Savings and Loans Societies had received loans totalling K158,486 while contributing savings of only K24,306 and the Registrar of the Federation, Girth McIlwain, said that the directors had seriously breached the rules for borrowing from their socieites. [29] The demise of the co-operative movement [30] may not have seen the end of the exploitation of villagers' savings within rural societies.

Class action: the control of political institutions

Participation in and attempts to control the political institutions established by the colonial and post-colonial state, local government councils, provincial government and national house of assembly, have been major activities of the rich rural classes. A fair degree of continuity of membership between co-operatives, councils and parliament was noticeable as the institutions were created in the 1960s. Fifteen of the thirty eight elected indigenous members of the house of assembly in 1966 had been, or actually were, associated with co-operatives [31] (a further indication of the reality of these organisations as rich peasant progress societies), and during the first house of assembly, 1964-68, 43 per cent of elected indigenous members had council experience. [32]

Local government councils were designed by the colonial state specifically to build upon customary social institutions and to be effective at the village level so that they would maintain 'tribal discipline'. [33] Councils without 'big man' support in the Eastern Highlands therefore tended to be rather ineffective [34] while the rich peasantry, firmly based as they were within the old societies, acted within and emerged out of their control of the councils. In the 1960s, five of Finney's ten Gorokan business leaders had become councillors of whom four had served either as president or vice-president of their councils, suggesting that when rising farmer-traders became councillors, they sought and attained the positions of power within them. [35] As this was done, the ambitious, successful farmer moved further upwards via the stepping stone of the council system to the national parliament or the provincial government (and its predecessor the Area Authority). The greater attractiveness of the parliament was reflected in the fact that three of Finney's ten men were Assembly members in 1968 and four more had contested national elections which were, in effect, thought Finney, 'contests between rival business leaders'. [36] The rising rich rural classes of the Eastern Highlands have dominated the region's parliamentary representation. This is clear from the results of the four general elections held to 1982 (ignoring white representatives in all cases). [37]

At the 1964 national elections two of the three elected members were rising rich farmer-traders:

1964 elections

Successful candidates	Ownership/activities
Sinake Giregire	Described above.
Muriso Warebu	Interpreter, councillor, 8,000 coffeetrees, timber mill (later minister in Eastern Highlands provincial government).
Ugi Biritu	Interpreter; peanuts, small amount of coffee*

In 1968, four out of five, conservatively estimated, were rising rich farmers:

1968 elections

Successful candidates	Ownership/activities
Giregire	As for 1964
Warebu	As for 1964
Sabumei Kofikai	Councillor; some 2,000 coffee trees and a trade store.
Bono Azanifa	luluai, council vice-president; some 5,000 coffee trees and a trade store.
Papuna Aruno	catechist; truck owner.*

In 1972, at least five out of seven elected members were rising businessmen:

1972 elections

Successful candidates	Ownership/activities
Giregire	As for 1964
Sasakila Sana (or Moses)	Some 200 acres with cattle
John Pokia	Obtained freehold over 47 hectares in Okapa in 1971; council president; cattle farmer, tradestore owner, health education officer (see chapter 4)
Akepa Miakwe	Council president; large coffee farmer, subsequently director of Bena rural development corporation as described above.
Suinavi Otio	Trucking business.
Atiheme Kimi	Interpreter, councillor.*
Onamauta Beibe	Interpreter.*

In 1977, in similar fashion, at least six out of eight were of the rich rural classes:

1977 elections

Successful candidates	Ownership/activities
Miakwe	As for 1972
Otio	As for 1972
Sailas Atopare	Former DASF employee, coffee farmer, trader, shareholder, director of a rural development corporation as described above.
Yubiti Yulaki	Coffee and cattle farmer, PMV operator.
Sunuki Kroki'e	Council president, member of interim provincial government; coffee grower and trade store owner.
Gai Duwabane	Coffee, commercial vegetables, shareholder, director of a rural development corporation.
Billy Hai	Coffee buyer.*
Undapmaina Kalagune	Council president, member of interim provincial government* (died December, 1978)

* Information on these candidates is scant. They have not been described as belonging to the rich rural classes for that reason.

Finney noted that each of the candidates in the 1964 election had a local tribal or regional base from which he could draw most of the votes cast, [38] and there is little reason to think that ethnicity and regionalism were not also important at subsequent elections. More recently, the Eastern Highlands' two leading rural development corporations, both of which are led by particularly successful members of the rich rural classes, are both locality based. Gouna in Goroka, Asaro and Unggai and Bena in the census division of that name but also reaching out, interestingly, to Kainantu at the end of the 1970s, and the third development corporation with Silas Atopare and Gai Duwabane among its directors, is based in the Asaro Watabung district. But it is not the case that ethnicity and class are forces working in opposite directions, rather that they are interwoven and that ethnicity offers support for the action of the rich peasantry and rural capitalists, whether in the socioeconomic form of the appropriation and exploitation of clan land, labour and savings, or politically for votes at elections. Ethnicity is a force of fluidity as well as of strength, and like so-called traditionalism it is moulded and changed by (and moulds and changes) the shape and appearance of the developing classes. [39]

The formation of the rich peasantry has partly derived out of the ability of powerful and ambitious individuals at one and the same time to manipulate ethnicity while transcending the limitations it might impose. The post war trajectory of political organisation in the Eastern Highlands has been toward bigger and bigger units. This was initially the case when villages grew in size following pacification; informal courts then extended the power and influence of indigenous leaders, a process formalised when local councils were introduced in the 1960s. Successful candidates were those who were able to secure a 'home base' in clan, tribal, village terms but then to move beyond it, through membership in or association with an organisation that transcended that immediate unit. It is not surprising then that successful candidates have had a close association with churches, or more recently with local councils, the provincial government and the development corporations. The winning individual was the man able to define his 'place' as larger than that of his opponents. This was only possible for those who possessed considerable wealth.

Commercial success was 'the outstanding accomplishment' of the Gorokans and others elected to parliament in the 1960s, [40] and was clearly a major feature in the success of, for example, Sinake Giregire and Muriso Warebu, later too of Akepa Miakwe. When Giregire was defeated it was not of course by a poor man or one of thousands of average coffee growers relying simply on the votes of his relatives and allies, but by a young rising rich farmer like Gai Duwabane whose development corporation had appeal and membership across the electorate. [41] The Gouna Corporation's Paul Ine (additionally president of the Goroka chamber of commerce and member of the interim provincial government) said directly that 'money is a tool for becoming a good leader'. Significantly, the Goroka chamber of commerce, through its vice-president, Paddy Leahy, 'pledged its support' to Atopare, Duwabane and Miakwe following their electoral success in 1977. [42]

The interim provincial government came into existence as the main provincial level political organisation in the Eastern Highlands in October 1976, superseding the Area Authority but retaining until the first provincial elections were held, in November-December 1978, the seventeen local council representatives as its core membership.

The following information concerns the class position of a sample [43] of sixteen interim provincial government members.

All sixteen members of the sample were coffee growers, thirteen owning more than 1,000 trees. Four members owned 3,000 trees, one five, one six and two 8,000 trees. Eleven owned pulpers, and two owned two pulpers. Ten of the sample members employed wage workers on their coffee farms, some employing as many as 15 during the coffee flush. Two had a permanent workforce of less than ten, the remainder having a permanent workforce of five or less.

Eight of the sample were engaged in other forms of commercial agriculture (cattle and commercial vegetable growing), one owning 100 cattle. Six had land registered in their own name; in one case 200 acres of agricultural land was registered, in another 100 acres. In two other cases small but valuable blocks of town land for commercial activities had been acquired. There was also a large involvement in trading capitalism: eight of the sample owned trade stores, one owning three and another four, and five were shareholders in substantial businesses.

This substantial involvement in agricultural and trading capitalism on the part of most members, meant that substantial incomes accrued to them. It is probable that K10,000 per annum represented a conservative average income with some members receiving incomes in excess of K20,000. A factor contributing to the relatively high economic position of members was the increased salaries announced in February 1978 for interim provincial government members.

Table 5.1
Annual incomes accruing to area authority/IPG members from their office, 1976, 1978

	1976 (K)	1978 (K)
President/Premier	960	5278
Vice-President/Deputy Premier	720	3978
Executive Committee/Ministers, Speaker	600	3484
Deputy Speaker	*	3276
Member	480	2496

Source: Donaldson and Good, 'Class, Power and Provincial Government in the Eastern Highlands', paper for the Waigani Seminar, UPNG, 1978; and Post Courier, 28 February, 1978.
* Not existing under the area authority.

In addition, the members continued to receive K7 per diem meeting allowance. The premier, deputy premier, the speaker and the ministers therefore enjoyed a 550 per cent increase in their official salaries over just two years, and ordinary members barely less. Among those in this fortunate position who have been referred to above were James Yanepa as premier; Ono Aia as deputy speaker; Muriso Warebu as a minister; Iyape Noruka, deputy premier and Hoffman Aize was speaker. These salaries rather vividly suggest not only that members of the rich rural classes have developed substantial common interests but that provincial government is, in effect, a means for protecting and furthering that common class interest.

Interim provincial government members made use of their newly acquired institutional position as a possible stepping stone to a higher political level during the 1977 general elections. Of the seventeen members who held office by representing local government councils, seven contested the general election. That these seven men were actively involved in provincial political affairs is clear from the fact that six of them sat on a committee of eleven set up in March 1975 to establish the interim provincial government. The seven contested five of the nine electorates and scored two wins and three seconds. Despite the handicap that in two electorates they campaigned against each other, the overall outcome was significant in itself and superior to some of the major political parties. [44]

A factor not necessarily unconnected with the electoral success of provincial parliamentarians is that the provincial government controls large financial resources. At its inception it received an interest-freeloan of K100,000 from the central government, a first annual capitation grant of K134,575, and a Derivation Grant of K62,000 for produce exported from the province in 1974-75. It controlled the allocation of Rural Improvement Funds to the province's eight local government councils amounting to K380,000 in 1976-77. It possessed certain taxing and licensing powers, and a new levy on alcohol consumed in the province, introduced in June 1977, was intended to increase revenue by K100,000 in the first eight months of its operation. [45] When the Eastern Highlands' first provincial budget was announced in February 1978, it totalled K2.3 million, and in January and in July of that year it received unconditional grants for the payment of functions assumed from the central government amounting to some K6 million. [46]

The provincial government's shareholding in Pacific Helicopters is indicative of its general orientation towards capitalist development. An area authority meeting, in December 1975, resolved that the authority 'does not want politics to interfere with development'. The authority, like provincial government after it, saw its role as promoting 'development' by assisting local businessmen, and in January 1975, the authority rejected a plan for the formation of a highlands bus company to be owned by the highlands area authorities 'because it could take money away from local businessmen'. In July that year, as well as investing money in Gouna's Pacific Helicopters Pty. Limited, the authority passed a resolution stating that it 'fully supported the company and agreed to write a letter to the Chairman of the Supply and Tenders Board asking that he give special consideration to Pacific Helicopters Pty. Ltd. in any application for work in or including the Eastern Highlands District'.

In February 1978 the premier of the provincial government reiterated similar sentiments when he complained that national government contracts for work within the province were going to businessmen from outside the province. James Yanepa called on the government to ensure that provincial businessmen were given priority in the allocation of government tenders for work in the Eastern Highlands.

The most important element in the provincial government in relation to the economic development of the Eastern Highlands and to the encouragement of capitalism, was Nokondi its 'business arm'. Nokondi was begun under the area authority, and its organisation and assets of some K140,000 were transferred to the interim provincial government. A large part of the conditional grants obtained from the national government were directed into Nokondi, in addition to the acquisition of share capital which, in July 1978, stood at K270,000. [47]

Nokondi in 1978 had K16,000 worth of shares in the once expatriate-dominated Eastern Highlands Farmers and Settlers Association Co-operative. It also owned and ran Kotuni commercial trout farm, had a forty per cent share of Goroka's leading hotel, had K40,000 invested in Pacific Helicopters Pty. Limited (a Gouna Development Corporation subsidiary) and the same amount invested in the Gouna Development Corporation itself.

The business arm with its substantial funds was controlled by a board of four directors which, in 1978, comprised Paul Ine as chairman, an expatriate on short-term contract as secretary, and the provincial deputy speaker and the agricultural minister. Ine and the secretary were known to be the most influential of the directors partly because of their education and superior understanding of company operations.

Nokondi has aimed at supporting development in the province and at generating a profit for itself. The latter goal has seen it buying two blocks of real estate in Goroka, as well as moving out of the province into the purchase of a shopping arcade in Papua New Guinea's capital Port Moresby. The acquisition of a seventeen per cent shareholding in Pacific Helicopters of Goroka, a twenty five per cent stake in the Gouna corporation, and a sixteen per cent share of the Goroka Softdrink Company, contribute both to commercial development in the province and its own profitability. It is notable that Harry Gotoha and Auwo Ketauwo had respectively, in 1978, a ten per cent and a four per cent personal holding in Pacific Helicopters, and that in the softdrink company, Gouna corporation had fifteen per cent of the shares and Harry Gotoha himself had thirty per cent. [48] While Nokondi planned to manufacture plastic containers possibly in Lae in association with Australian intersts, its existing investments were interesting because of the links that had been established with Gouna corporation and with its leading directors Gotoha, Ketauwo, and of necessity Paul Ine who, in 1978 as noted, was simultaneously an executive director of Gouna and chairman of Nokondi.

Association with Nokondi is undoubtedly of great value to Gouna and to these individual businessmen because of the resources which it commands through the governments in Goroka and Port Moresby. [49] Nor do these linkages seem likely to be broken in the near future, for as a result of the last provincial elections of the decade, despite Ine's retirement, two executive directors of Gouna were elected to the provincial assembly, Isembo Olekevea and Auwo Ketauwo. Subsequently Ketauwo was narrowly beaten for the premiership, 16 votes to 13, by the incumbent James Yanepa. [50] Nokondi planned eventually to sell back its shareholdings to Gouna and then repeat the process with other local companies, presumably an acceptable strategy to businessmen in receipt of such investment funds.

Members of the rich rural classes have attempted to use their position in provincial and national assemblies to promote their common interests in a number of key areas, directly through new or amended legislation and indirectly in public debate utilising their status as elected leaders. They have attempted to restrict the sale of unprocessed coffee on the roadside, they have avoided the imposition of a government export tax on the commodity, and they have supported the continuance of the plantation system and the marketing of coffee through private ownership. [51]

Control over land, as already suggested, has been a key concern of the rich rural classes, and the area authority in 1974, passed resolutions asking the Director of Lands, Survey and Mines to find ways to speed up and simplify land acquisition and leasing procedures, and that the authority be given power to initiate forfeiture action with respect to undeveloped leases. Premier James Yanepa stated in December 1976, that existing land laws 'hinder the more active people who wish to get involved in business', and he called for amendments which would 'allow different groups of Papua New Guineans to buy properties in other areas or provinces'. [52]

Similarly, rural wages have been an area of prime concern, and both individual rich farmers, such as Sinake Giregire and John Akunai, and a sub-committee of the area authority have appeared before the Minimum Wages Board to argue against any increase in wages. Opposition to the working class was also expressed within the house of assembly in 1975, defeating a bill requiring employers to deduct union fees from employees' wages, which promised to strengthen the trade union movement in the country.

Resistance to the rise of the rich rural classes has taken a number of forms, among them a recrudesence of sorcery and an increase in the frequency and magnitude of coffee stealing. Sorcery had been discussed at some length in the national parliament and in the area authority, and coffee stealing was referred to by Premier Yanepa in 1977 when he demanded that 'the police must be tougher and the penalties must be stronger'. Yanepa threatened to introduce by-laws allowing the police to be armed 'to stop this uncontrolled stealing'. His sentiments were echoed by Akepa Miakwe who said that the Bena development corporation had 'terrible trouble' on its plantations in 1976 and that the situation had worsened in 1977. [53] In March 1978 plantation owners met at the Asaro Coffee Estate near Goroka and called on the national and provincial governments to introduce legislation which would ensure that people convicted of stealing receive minimum jail terms of six months. [54]

This posture of rectitude towards the lawlessness of the lower orders had not prevented the representatives of the rich rural classes from strongly opposing the application of a code of ethical behaviour on their own business activities. The proposed national Leadership Code of 1978 was criticised by the interim provincial government. Premier Yanepa said that 'a man's ability as a businessman ... was in jeopardy', and he declared that 'Highlands politicians and officials would rather resign than sell their business interests'. [55] This was a critical event and there is little doubt that the provincial government accurately expressed the attitude of other rich farmers of the highlands. Not only therefore has the provincial government, through Nokondi, acted to give considerable financial aid to prominent businessmen in the Eastern Highlands, but it has also asserted itself nationally to obtain a better deal for the province's rich rural classes, and it has endeavoured to protect their property and weaken the position of rural workers. Its members have also feathered their own nests very nicely.

Class action: the role of the development corporations

The organisation and control of the development corporations represents further significant action by leading elements of the rich rural classes. Yet it is characteristic of these organisations that they possess a populist image and are seen to work in the interests of all members of the community. John Kaputin said of the New Guinea Development Corporation of the Gazelle Peninsula, of which he was chairman of directors, that 'the corporation has done a great deal to develop and promote the concept of group participation in economic development throughout the country'. [56] An anonymous correspondent from Goroka ('village boi') said, in addition, that:

> village-based groups such as the Matanguan Association and the Gouna Development Group ... show more meaning and are more effective [than the political parties] when it comes to actual village development. Here the individual is moved because he sees the benefits coming and this stimulates him to contribute his efforts and resources to the development initiated at his level. [57]

Not only is this true to a limited extent, but the two main development corporations of the Eastern Highlands were also through the 1970s notably successful in a commercial sense. The Bena corporation had acquired ownership of four coffee plantations including, on the Urona estate, one of the largest processing factories in the country, and it subsequently spent close to K200,000 on improvements to this concern. In addition, it possessed cattle at Kainantu and a shopping complex in Goroka. Gouna, as already noted, owned some 260 acres of coffee and in 1977 its total fixed assets were valued at K700,000. [58] But Gouna also had wide commercial involvements reflecting the big ambitions of some of its directors,and its success was well enough indicated by the performance of Pacific Helicopters, an ambitious operation, since its formation in April 1975. The fleet of helicopters had grown from one to four by 1979, permanent servicing and engineering facilities have been established in Goroka and, not least importantly, national ownership had increased from 40 per cent to 51 per cent in less than four years, with total ownership and management by indigenous businessmen intended in the near future. Later the company won an exclusive and lucrative hiring rights contract from the national government's Postal and Telecommunications Department, and planned (successfully, as will be seen below) to increase its helicopter fleet further. [59]

The performance of Gouna and Bena is greatly superior to that of the co-operatives earlier, and to an individual rich farmer like Sinake Giregire who has acted alone entrepeneurially and not achieved all his expectations of the 1960s. While the co-operatives were sometimes started for DASF's own policy ends and mismanagement by rising rich peasants was occasionally conspicuous, today's development corporations employed management advisers on short-term contracts [60] and displayed efficiency. Both Gouna and Bena represented the large scale private ownership by indigenous people of land, crops, equipment and commercial resources.

Another characteristic of both corporations distinguishing them clearly from the co-operatives and, as 'village boi' accurately noted, from political parties, is the large memberships and the sizable investment funds they have mobilised from villagers. Bena in 1977 had a membership of approximately 15,500 people and from its members of two years' earlier it raised some K50,000 with which it obtained both a government and a bank loan and purchased its plantations and processing works. Gouna had 2-3,000 'shareholders' which were mainly organised in village groups but also included Nokondi. The villagers contributed about K125,000 with another K40,000 coming directly from the provincial government 'business arm'. [61] Bena had its village members organised into small 'business groups', and each group (fifteen altogether in 1978) bought shares from the corporation and was represented within the organisation 'on a committee basis'. [62] Gouna in 1977, had its membership organised in eight business or village groups (including Nokondi), and each group 'had links with' one member of a nine member board of directors. With its diversified operations, the Corporation was divided into seven different subsidiary companies, Gouna Investment, Gouna Plantations, Properties, Trading, etc., to minimise financial risks and supposedly to simplify management and accounting for the group as a whole. [63] Finally, both development corporations also represented the large scale mobilisation of wage labour by nationals, Bena particularly so, as noted above, given its greater concentration upon coffee production.

Gouna and Bena have succeeded in mobilising villagers' savings and labour while espousing an ideology of developmentalism and populism. This ideology has clearly appealed to the people of the Eastern Highlands, but it has also served to mask the reality of control of the corporations by rural capitalists. Akepa Miakwe founded the Bena corporation in 1973, because he believed that the coming of independence meant opportunities for local businessmen to purchase companies and plantations from their departing foreign owners. [64] Hari Gotoha started Gouna at around the same time, after thinking about 'how europeans start businesses', and with a similar aim of gradually gaining control over the foreign owned companies in the region. [65] The expansion of both founders' personal assets and the development of their respective corporations subsequently went on together, while each, in company with a small number of close associates and business partners, retained firm control over their organisations. In the case of Gouna, an individual shareholder is, in general, dependent upon the good relations of his or her business group with one of the nine directors if he or she is to have any influence within the organisation. This is supplemented by meetings of up to 1,000 people which are occasionally held to explain policy, but here again the illiterate villager might be at some disadvantage. In any case, not all of the eight business groups are equal in their relationship with the board of directors for one of them is Nokondi. And on the board of directors itself, three or four closely associated men wielded decisive influence. Gotoha, Secretary of the board, personally collected the money from villagers investing in the corporation and kept group deposit books in his house. [66] The growth of Gouna's corporate assets since 1973 has been accompanied by the commercial rise of the rural capitalists which formed and control it. Bena Corporation and Akepa Miakwe appeared similarly interrelated.

Control of the development corporations has also led to significant political action at the national and provincial levels. When only two sitting indigenous members of the national parliament from the Eastern Highlands were successful in winning re-election in 1977, one of these was Akepa Miakwe, and among the new members both Sailas Atopare and Gai Duwabane were senior figures in the Asaro Watabung Rural Development Corporation. [67] Miakwe subsequently became opposition shadow minister for Lands, and Atopare shadow minister for Transport and, following the establishment of the Pangu-United Party coalition government in 1978, Duwabane was named minister for Defence.

The political power of executive position in the development corporations was further demonstrated at the elections to the Eastern Highlands provincial assembly in December 1978 when, as indicated, both Ketauwo and Olekavea won seats from the Goroka area. [68] This increase in Gouna's political strength would place it in a better position to deal with 'interference in business' by Nokondi, a problem which Gouna and Bena have each complained of, the latter more vehemently. [69] Each of the two corporations want Nokondi to be involved in business only on a limited scale, certainly on a minority basis in association with established business, and preferably in support of their own corporate and individual commercial activities. [70] Gouna, of course, had already been assisted by Nokondi in precisely this way. Nevertheless Nokondi's future role remained slightly uncertain given the financial and other resources which might be available to it through the national and provincial governments.

Relations between the two main development corporations were apparently cordial. Akepa Miakwe considered that Bena and Gouna have agreed, so to speak, to a division of labour, where Bena would remain in coffee while Gouna would concentrate on the acquisition and development of commercial property. [71] This strategy had however been modified by Gouna's comparatively modest move into coffee and Bena's into Gorokan real estate. Gouna accepted the general thrust of the supposed agreement, partly because Gotoha considered the coffee market too unreliable, but Ketauwo spoke of another informal agreement whereby remaining foreign owned plantations on the Asaro side of the Highlands Highway would go to Gouna in Goroka, Unggai and Asaro. Gotoha and Ketauwo were shareholders in Bena, and Bena lent Gouna K110,000 for its purchase of Jim Leahy's plantation. [72] Gouna and Bena's relations were therefore those of friendly competition, but given changing economiccircumstances, the maintenance of cordiality was perhaps dependent upon the investment opportunities, the ambitions and resources of the men controlling the two corporations and, of course, the role of Nokondi, the only comparable indigenously owned business organisation in the province. [73] In the meantime, the development corporations together and separately, continued with their action in the cause of rural capitalism, class action which is the more effective precisely because it successfully utilised localism, regionalism and populism.

The reactions of the poor peasantry and the limits of class action

The growth of agricultural and trading capitalism and the rise of the rich rural classes have been twin aspects of the one process, the implantation of a capitalist mode itself, limited by the near absence of industrialisation and still rooted in some crucial institutions originating from pre-capitalist societies. Such transformation and conservation over little more than thirty years was bound to bring in its wake deep problems for the mass of the peasantry who bore the burdens of the gross inequalities in control over and access to land, coffee trees, cattle, labour, equipment, extension services, loans and incomes which were integral to the new production system. Many indications of a deepening malaise have appeared. The concentration on cash crops for export had a detrimental effect on domestic food production and contributed to the low health status of Eastern Highlanders. Even the enjoyment of high coffee prices has not led to any obvious amelioration of the situation for most small farmers in the province but rather, on the experience of 1976, to rapidly rising expenditures on imported, replaceable food items and to lavish outlays on beer. In June and July 1976 beer sales reached K800,000 or roughly K3 for every man, woman and child in the province. [74]

Deep social unrest is implicit in the lawlessness on the main artery of the region, the Highlands Highway, down which all coffee travels to the sea and up which imported products enter the highlands. The expatriate Goroka based principal highlands trading firm of Collins and Leahy estimated in 1977 that the company was losing an average of K100 every time a truck was pilfered, and the action was such as to require the use of police to escort every truck through the known trouble spots. [75] In the view of other commercial representatives, particularly trucking companies, the governments should 'consider the use of Army personnel to assist police in bringing under control the blatant disregard for law and order'. [76] The area between Kassam Pass and Goroka was particularly notorious with three 'major incidents' of theft from trucks occurring in one night, but the incidence of theft was widespread with all centres from Lae to Hagen regularly receiving reports of the slashing of tarpaulins and assaults on drivers. [77] The level of violence and theft on the highway was reflected in the experience of Soso Subi, one of the earliest of rich coffee growers (see chapter 4) who, fearing trouble on the Highlands Highway when travelling on 10 March 1977, carried a loaded shotgun, and later safely at home accidently discharged the gun killing his teenage son. [78]

Violence associated with property was also represented in 'tribal fighting', much of which is actually concerned with land boundary disputes. In the early 1970s armed clashes over land in the Eastern Highlands were officially reported to occur 'very frequently'. [79] Near the end of the decade, fights in the highlands were reported to be increasing in number, size and duration, with a particular rise in the incidence of viciousness and destructiveness. [80] Most fighting took place in Simbu, Enga and the Western Highlands, but in the Asaro area of the Eastern Highlands fighting,involving loss of life, went on for more than a month in late 1978 and ended only after peace talks initiated by the provincial government. [81] It is evident in Simbu that clan leaders have used so-called traditional sanctions to manipulate landless and unemployed youths into taking a 'front line' role in warfare which is aimed at 'acquiring additional land or destroying the opposition economically'; the victorious group and their leaders 'thereby gain local prestige and political power among neighbouring clans'. [82] In June 1979, killings and the destruction of property and the loss of export earnings was such that the national government took the extreme step of declaring a State of Emergency in the highlands. According to the prime minister '... in the past six months of this year there have been 52 clan fights in the highlands involving 108 separate clashes which have resulted in 35 people being killed, 207 wounded, 467 houses, 6 trade stores and one workshop destroyed, sense- less killing of domestic animals and fowls and large scale theft of money and food' as well as the destruction of coffee trees and gardens. [83]

Coffee stealing in the Eastern Highlands is another socially significant activity. In April 1977 it was reported as out of control with, according to Paddy Leahy of the Asaro Coffee Estate, villagers walking on to plantations in daylight, stealing the coffee and selling it by the roadside, with losses approximating thousands of Kina a week. Asaro Estate therefore employed three full time armed guards and Akepa Miakwe said that the situation on the plantations owned by the Bena Corporation was even worse than the previous year. [84] About a week after Miakwe's statement, three watchmen were beaten up by intruders who climbed a two metre high security fence on the Nomba estate at north Asaro. Nomba was already experiencing two or three cases of coffee stealing each week, and the plantation manager attributed the problem to 'the young school dropout, aged about 18, who has no job, no responsibility and is generally frustrated. He consoles himself by stealing'. [85] Premier James Yanepa spoke for all plantation owners when he declared that 'law and order is nearly breaking down at the moment', and called for the police to get tougher and for penalties to be made more severe. [86] In August 1978, however, three tonnes of coffee (value approximately K4,000) were stolen from the Roka estate near Goroka, with some 65 men taking part in the robbery, the sixth at the plantation in that year. [87] The regular placement of armed guards on the big properties resulted directly from the perception of the larger growers that they stood to lose well over K1 million in 1977. Long standing coffee grower Biato Hero claimed that the major effect of such action was to force thieves to concentrate on smaller plantations like his own. [88]

Some of this activity, highway robbery, violent land disputes, coffee stealing, resembles social banditry in its scale, persistence and opposition to property and property owners. It would seem, in part, to represent inchoate class action by poorer peasants and perhaps the young unemployed who experience the heavy inequalities and now dwindling opportunities of rural capitalism. Land shortages have appeared and, as noted, young men are finding it very difficult to enter coffee farming. The members of the rich classes with striking unanimity, certainly perceive this action as a grave threat to themselves and their property, and react in their turn with high wire fences, armed guards, shotguns and calls for tougher police action, curfews, the use of the military and stiffer penalties.

However, no political organisations exist to represent and mobilise the poorer peasantry and the new working class for the achievement of a more equitable economic system. Trade unions have been formed in the two main towns of the Eastern Highlands, but they have difficulty in gaining and maintaining a membership in excess of 200 workers and are often beset by financial and other difficulties. The Goroka Workers Association was formed in the mid-1960s with a membership of almost 300, but this later declined to little over 100 despite the fact that the union was officially reported as 'reasonably active' at the start of the 1970s. The Kainantu Workers Association was established in 1970 but had gained less than 100 members four years later. [89]

Trade unions also face the strong hostility of the rich classes, especially on the key issue of wages. Sinake Giregire has pictured a possible increase in workers' wages and conditions as a blow against all village people and a barrier against further foreign investment, and others, such as Sailas Atopare in August 1978, were ready to defend foreign investment and existing economic arrangements against what they said were communist tendencies in national government policies. [90] The national political parties for their part, offer no alternative to the dominance of the rich rural classes, for they barely existed organisationally outside the brief effervescence of the general elections and then largely functioned as mechanisms to elevate likely candidates from the rich rural classes into parliament. [91]

The mass of peasants and workers in the Eastern Highlands are without organisational support but this does not fully define their weak situation. The continuing appeal of ethnicity and regionalism encourages the poorer classes to continue to see their interests as being one with the rich rural classes, as 'rubbish men' earlier did with the big men. Provincial government gives a new dimension to the spirit of regionalism protecting the dominant groups, and the main development corporations add the potent influence of developmentalism to the strength of some rich men in certain districts. An acceptance of the political power of the rich now, as with the big man of the past, is therefore part of the subordination of the poor, despite the fact that the increasing incidence of lawlessness, violence and robbery is simultaneously directed against the property of the rich.

Strong within their own province, the rich rural classes remained weak nationally and internationally. Both the rich peasantry and rural capitalists are dependent upon external sources for their propsperity and protection, the former particularly so. As with the colonial state earlier, both are reliant upon the policies and resources of the successor state in a wide variety of ways. Bank loans and agricultural extension aid are still required, though credit is now more readily available from the commercial banks to the successful indigenous business, and there were indications that the Development Bank was less active in commercial agriculture and the Department of Primary Industry's aid to coffee production in the province was minimal in the mid-1970s. The rich peasantry is the more affected by a reduction of DPI's activity than the rural capitalists who have diversified operations and, through their development corporations, can appoint their own extension workers or didimen. [92] Access to land, that is, to legal ownership over land, is a critical need for the rural capitalists, but the national government was considering new laws which promised to ease the situation considerably. The successor state also has shown sympathy to the rich rural classes with regard to wages, and it has turned a blind eye to the indigenous employer's widespread practice of paying workers below the legal minimum wage. The provision of new business opportunities is another area where state initiative is important, and the independent government has voluntarily withdrawn from certain activities in favour of rural traders. [93] Bigger potential opportunities have arisen, however, through the state's most positive initiative in establishing the system of provincial government. And the rural rich are also dependent on the police and the national courts to protect them from the lawlessness which their own accumulation promotes.

The extra-provincial power of the rich rural classes is gravely weakened by their lack of formal education and their profoundly peasant like ignorance of the outside world. Not only have the highlands been historically deprived of their share of educational resources, as noted the national educational status of the Eastern Highlands was second bottom, but the typical coffee farmer in particular was 'a middle aged man with no formal education'. [94] Rural capitalists are able to compensate for this disability to some extent through the employment of skilled managers in their development corporations and through close association with young, educated men. Kumoro Vira, for example, a former manager of the Eastern Highlands Savings and Loans Society, had advised Hari Gotoha on outside events and was described by Gotoha as 'like a son'. [95] But the rich rural classes remain in a weak position in the broad field of national and international politics, and this increases their reliance upon the post colonial state, or rather on the direct controllers of this institution, the educated petty bourgeoisie. [96]

This new administrative class had, with independence, a limited but real freedom to develop national policy and their choice, on options like a concentration or otherwise on minerals' exploitation or domestic food production, could materially affect the classes dominant in cash crop production for export. The power of the educated petty bourgeoisie is largely restricted to their position within the bureaucracy (supplemented perhaps by their international links with metropolitan bourgeoisies). The rich rural classes, on the other hand, have a strong position in village society and are established in a production system which earned important revenue for the country, more in fact than that which was earned by any other class or group of indigenous owners. Each offered something to the other in national politics, which had been characterised through the 1970s by a tacit and uncertain alliance between the two, unexpressed by the existing political parties and national ideology and affected by rivalries between the three geographical divisions in the country, the coast, highlands and islands. With the focus of this relationship heavily weighted towards the resources of the state, the initiative in day to day politics stood with the bureaucratic stratum. But the structures of commercial agriculture (in both production and marketing in coffee) are commanded increasingly by the rural rich and built upon pre-capitalist social institutions, and short of the most unlikely changes of policy, say a non-capitalist rural development programme, their established yet dependent position was secure. The rural poor, faced by such ruling class alignments, are left with few political alternatives, and their frustrations will therefore grow and they may increasingly tend to 'console themselves' with the banditry and incipient rebellion that has characterised the resistance of the mass of the peasantry in so many times and places.

The growth of an export-orientated commercial agriculture in PNG has been based on only two or three crops, and is highly concentrated geographically. Copra and cocoa have been chiefly located in the New Guinea islands and coastal mainland, and coffee in the highlands. Even within these regions, the most intense concentration on copra and cocoa was probably found in East New Britain and on the Gazelle Peninsula, and of coffee, over most of the period in the Western and Eastern Highlands.

Despite national and international dependency, the achievements of the rich rural classes of the Eastern Highlands are, however, particularly extensive. They steadily increased their hold over coffee production. They have a strong control over the growing and processing stages and were gaining control in marketing; [97] they formed and control the expanding successful development corporations; in transport, they have gone beyond the ownership of taxis and trucks and into aviation, and they are involved in manufacturing, actual or planned, in softdrinks, plastics, timber and clothing. This was not, of course, high level manufacturing activity, but it was, in company with coffee production, located mainly in the Eastern Highlands itself, rather than is usual in the urban enclaves of Port Moresby or Lae. By 1979 the Eastern Highlands contained 60 factories employing 1,800 workers and ranked fourth among the provinces in the number of factories operating and workers employed. In terms of salaries and wages paid per employee though, the Eastern Highlands Province was below the average for the highlands as a whole, and ranked seventh from the bottom nationally. [98] Some accumulation of capital is beginning to take place which has national as well as provincial significance.

Notes

[1] Finney estimated the total value of Hari Gotoha's property, part of which was held in a partnership with Soso Subi at $100,000. B.R. Finney, Big Men and Business, Canberra, ANU Press, pp. 85 and 104-106.

[2] Personal interview with Hari Gotoha, Goroka, 4 May 1977.

[3] Personal interview with Anders Berqvist, Gouna business manager, 4 July 1977. 'Gouna' is an acronym for Goroka, and adjacent Unggai and Asaro (GO/UN/A).

[4] Personal interview, Goroka, 5 July 1978.

[5] Interview by Oltai Phile with Auwo Ketauwo, Goroka, 10 November 1978.

[6] Interview by Oltai Phile with Isembo Olekavea, Goroka, 15 November 1978.

[7] Interview by Aita Ivarato with Paul Ine, Goroka, 27 June 1978; and Post Courier, 1 March 1979.

[8] Gouna corporation paid each of its directors an honorarium of K100

a month, and a number of directorships would therefore represent a worthwhile increment to the annual income of a national entrepreneur. However Gotoha is slightly different from his colleagues in his narrower concentration on business rather than political activity.

[9] Personal interviews, Goroka, 4 May 1977 and 6 July 1978.

[10] Personal interviews, Goroka, 14 and 15 September 1978.

[11] The use of the term unformed classes in colonial and post-colonial situations is noted in Bjorn Beckman, Organising the Farmers, Uppsala, Scandanavian Institute of African Studies, p. 23.

[12] Beckman designated four classes within cocoa production in Ghana: an upper layer of big farmers or farmer traders with a heavy reliance on hired labour, producing 200 loads of cocoa or more and with an important involvement in farmers' organisations; those middle farmers employing several labourers; those middle farmers with holdings requiring the labour of only one adult; and the lower stratum of smallholders producing less than 40 loads, ibid., p. 40.

[13] National Statistical Office, Rural Industries 1979, Port Moresby, 1982, Table 7.

[14] National Statistical Office, Summary of Statistics - 1978, Port Moresby, 1981, p. 50; and Rural Industries, ibid., p. 3.

[15] The health status index is based on rural life expectancy, rural child mortality, and malnutrition level; only the Sepik provinces ranked lower than the Highlands. The education status relates to adult literacy, and the percentages of 7 year old children enrolled in Grade 1 and of 13 years olds in Grade 7. National Public Expenditure Plan 1978-1981, Exhibits 3.1 and 3.3.

[16] Finney, op. cit., p. 179.

[17] Reflected in the wide public discussion among leading political and religious figures at the time, and noted in Good, 'Formation of the Peasantry', in Amarshi, Good and Mortimer, Development and Dependency: The Political Economy of Papua New Guinea, Melbourne, Oxford University Press, 1979.

[18] Minutes of Directors' meeting, 12 January and 14 June 1965, and DASF Report, March 1965, file 23-3-N(2), DPI Archives, Goroka.

[19] DASF Report April-May 1965, and Directors' meeting, 1 January 1966 and 4 October 1966, DPI Archives, Goroka.

[20] Minutes of Directors' meeting, 3 August and 14 December 1965; and 1 November 1966 and 11 April 1967, DPI Archives, Goroka.

[21] Minutes of General Meeting and Directors' meeting, 5 July 1966, and letter from the Registrar of Co-operative Societies, 18 August 1966, DPI Archives, Goroka.

[22] Minutes of Directors' meeting, 13 June 1967, and DASF Report 21 February 1971, DPI Archives, Goroka.

[23] Letter from J.A. Langry, co-operative officer, Goroka 29 July 1971, file 23-3-N(2), DPI Archives, Goroka.

[24] Letter from the acting assistant director, Development and Marketing, 24 June 1971, file 23-3-M(3), DPI Archives, Goroka.

[25] Letter from the district rural development officer, Goroka, 3 February 1969, and minutes of General Meeting, 31 July 1970, DPI Archives, Goroka.

[26] Minutes of General Meeting, 10 October 1969, file 23-3-N(3), DPI Archives, Goroka.

[27] Personal interview with Kumoro Vira, Goroka, 28 April 1977. Vira was the former manager of the Eastern Highlands Savings and Loans Society, a close business associate of Gotoha and Ketauwo, and

a candidate for parliamentary election; and Post-Courier, 25 October 1977.

[28] Personal interview, Goroka, 5 July 1978.

[29] 'Business and Finance Summary', Office of Information, Konedobu, 7 April 1978.

[30] Further on co-operatives in the Eastern Highlands, see Donaldson and Good, 'The Eastern Highlands, Coffee and Class' in Denoon and Snowden (eds), A Time to Plant and a Time to Uproot: A History of Agriculture in Papua New Guinea, Institute of Papua New Guinea Studies Port Moresby, 1980, pp. 152-6.

[31] H.H. Jackman, 'Co-operatives', Encyclopedia of Papua New Guinea, Clayton, Melbourne University Press, 1973.

[32] Peter Hastings, New Guinea: Problems and Prospects, Melbourne, Cheshire, 1973, p. 125.

[33] In the words of D.M. Fenbury, architect of the Local Government plan, quoted in David Simpson, 'The Administration and Political Development of Local Government in Papua New Guinea to 1946', unpublished BA Honours thesis, Department of Political Studies, UPNG, 1976, pp. 13-14.

[34] See L.L. Langness, 'Notes on the Bena Council, Eastern Highlands', Oceania, 23, 3, 1963.

[35] Finney, op. cit., p. 117.

[36] Ibid., pp. 116 and 119. The council system remained important not least because it acted as a conduit for the national government's Rural Improvement Funds. Of a sample of 45 candidates at the 1977 general elections in the Eastern Highlands, 15 were council members and of those candidates who were classifiable as either rich peasants or rural capitalists, 64 per cent were councillors. Mike Donaldson and Kenneth Good, 'Class and Politics in the Eastern Highlands of PNG', History of Agriculture Discussion Paper No. 9, UPNG, March 1978a, pp. 20-21.

[37] Main sources of brief notes on candidates' property and activities are biographical details published by the house of assembly for the Eastern Highlands, and for the personalities of the sixties, Finney, op. cit. The data is indicative only and not comprehensive as regards a political career.

[38] Finney, op. cit.

[39] J.R. Gusfield, 'Tradition and Modernity: Misplaced Polarities in the Study of Social Change', American Journal of Sociology, 72, 4, 1967, p. 358.

[40] Finney, op. cit., p. 118.

[41] The success of candidates from the rich rural classes in the 1977 election is discussed in Donaldson and Good, 1978a, op. cit.

[42] Interview by Aita Ivarato with Paul Ine, Goroka, 27 June 1978; and Post-Courier, 19 March 1977.

[43] Membership of the IPG was twenty three which included representatives of women, the church and the Gorokan Chamber of Commerce (Paul Ine). For a discussion of the sample and of members' characteristics see Donaldson and Good, 'Class, Power and Provincial Government in the Eastern Highlands', paper for the Waigani Seminar, UPNG, 1978b.

[44] Considered further in Donaldson and Good, 1978a, op. cit.

[45] Donaldson and Good, 1978b, op. cit., and Post Courier, 15 February and 30 June 1977.

[46] Post-Courier, 13 February 1978, and personal communication, Department of Finance, 12 March 1979.

[47] Details of Nokondi mostly derive from a personal interview with

John Peterson, Business Development Office, Goroka, 7 July 1978.

[48] Personal interview with Anders Berqvist, Goroka, 4 July 1977.

[49] It is also noteworthy that Barry Holloway was the (foreign born) member for the Eastern Highlands Provincial seat in the National Parliament and, after the 1977 elections, was the country's finance minister as well. He was known to have advised Nokondi that the aforementioned shopping arcade in Port Moresby was available for purchase.

[50] Ketauwo afterwards declined the offer of the provincial finance portfolio and declared that he would lead the opposition within the assembly, Minutes of the Provincial Assembly Meeting, 21 December 1978.

[51] See for example the reaction to the purchase of the largest coffee exporter, ANGCO, by the PNG government-owned Investment Corporation, followed by assurances from Prime Minister Michael Somare of support for the private sector in coffee marketing. Post Courier, 28 March 1977.

[52] Donaldson and Good, 1978a, op. cit.

[53] Post-Courier, 27 April 1977.

[54] Post-Courier, 9 March 1978.

[55] Post-Courier, 7 March 1978.

[56] Post-Courier, 16 June 1977.

[57] Post-Courier, 17 June 1977.

[58] Donaldson and Good, 1978a, op. cit.

[59] Post-Courier, 9 February 1979. It will be recalled that Gotoha was the largest indigenous shareholder in the company, and that Ketauwo and Olekavea were also involved, along with Goroka Tyre and Battery which belonged to Gouna and to Gotoha personally.

[60] Bena's position was a little different in that its management services were supplied under an agreement by the large (later government-owned) marketing organisation, ANGCO.

[61] Donaldson and Good, 1978a, op. cit. These figures compare favourably with the earlier work of the New Guinea Development Corporation which, according to John Kaputin, raised K125,000 in two and a half years. Post-Courier, 31 August 1977.

[62] Personal interview with Akepa Miakwe, Goroka, 6 July 1978.

[63] Personal interview with Anders Berqvist, Goroka, 4 July 1977.

[64] Post-Courier, 19 March 1977, and personal interview with Akepa Miakwe, 6 July 1978.

[65] Personal interview with Hari Gotoha, 5 May 1977.

[66] Ibid.

[67] This is a somewhat different organisation from Gouna and Bena; it was originally Fointa coffee plantation, owned by Fred Leahy who later acted as paid manager of the corporation, and it used the profits from the plantation to purchase tractors and other equipment to make available to the people of the locality. It had no members or shareholders in the Gouna and Bena sense, but it aimed at profit making, espoused developmentalism and populism and had an indigenous leadership similar to Gouna and Bena in all but their years.

[68] Both actively appealed to voters as leaders of Gouna Development Corporation. Interview by Oltai Phile with Ketauwo, Goroka, 10 November and with Olekavea on 15 November 1978. Ketauwo ran in the Gahuku constituency which surrounds Goroka, where he defeated the sitting provincial government speaker, Hoffman Aize, and Olekavea was successful in Mimanalo constituency which borders on Gahuku to the north west.

[69] According to Akepa Miakwe, Bena had wished to purchase Goroka's largest hotel, the Bird of Paradise, but Nokondi had 'interferred' preventing this; Nokondi is also 'confused' as to its role, and is dominated by politicians. Personal interviews 4 May 1977 and 6 July 1978.

[70] Personal interviews with Gotoha and with Ketauwo, 5 July 1978.

[71] Personal interview with Miakwe, 4 May 1977.

[72] Personal interviews 5 May 1977 and 5 July 1978.

[73] The Lowa Development Corporation (later, the Lowa Investment Corporation) was established in 1970, involving a number of villages east of Goroka, to assist them to market fresh food and vegetables. Capital expenditure had exceeded K10,000 by 1973, but in 1974 the government opened a fresh food agency in Goroka, and in December 1975 Lowa closed own, disposing of its trucks, and renting its warehouse to its state-owned competitor. In October 1976, the Lowa Investment Corporation was formed to invest in a number of enterprises in and around Goroka. It paid its first (ever) dividend of 20 per cent in December 1976. Post-Courier, 1 February 1977.

[74] Post-Courier, 30 September and 20 October 1976. See J.N. Lambert, 'Does Cash Cropping Cause Malnutrition?'. National Planning Office, 1978, mimeo; and Stewart MacPherson, 'Health and Underdevelopment in Papua New Guinea'. Paper presented to the Congress of the Australian and New Zealand Association for the Advancement of Science, January 1979.

[75] Post-Courier, 20 and 23 May 1977.

[76] Post-Courier, 16 May 1977.

[77] Ibid., and 23 May 1977.

[78] He was found guilty of criminal negligence in the national court in Goroka. Post-Courier, 16 June 1977.

[79] Department of the Chief Minister, Eastern Highlands District, 'Information Brief', February 1974, p. 13.

[80] Post-Courier, 15 August 1977.

[81] Post-Courier, 14 September 1978. The provincial government does not always play such a mediating role. 'Tribal fighting' in Goroka township in August 1982 involved an assault on 3 provincial government officials. Illawarra Mercury, 1 September 1982, p. 15.

[82] Economic Consultants Limited in association with Crown Agents for Overseas Governments and Administrations, Simbu Province: Rural Development Study, Interim Report. Kundiawa, 1 March 1979, p. 131.

[83] Post-Courier, 23 July 1979.

[84] Post-Courier, 27 April 1977.

[85] Post-Courier, 3 May 1977.

[86] Post-Courier, 27 April 1977.

[87] Post-Courier, 16 August 1978.

[88] Post-Courier, 31 May 1977.

[89] Fitzpatrick and Good, 'The Formation of the Working Class' in Amarshi, Good and Mortimer, op. cit., and Annual Report, Eastern Highlands District, 1970-71.

[90] Post-Courier, 12 March 1976 and 31 August 1978.

[91] For example, David Hegarty, 'The Political Parties'in Amarshi, Good and Mortimer, op. cit.

[92] Finance Minister Barry Holloway said that the share of loans going to agriculture seemed to be declining, Post-Courier, 3 November 1978; and Miakwe, Ketauwo, et. al. were critical of what they

termed the 'small mindedness' of DPI policy.

[93] For example, a government preference for private trucking contractors rather than the continued use of government vehicles was announced. 'The Formation of the Peasantry', op. cit. The state's tax concessions to national businesses are also important.

[94] D. Anderson, An Economic Survey of Smallholder Coffee Producers - 1976, Department of Primary Industry, Port Moresby, September 1977.

[95] Personal interviews 5 May and 4 July 1977. Sailas Atopare was unusually a young and educated man.

[96] The Eastern Highlands national parliamentarian Suinaivi Otio said that less educated parliamentarians like himself would be only too willing to accept advice from the educated elite, and the deputy premier of the interim provincial government referred in 1976 to the difficulties of drawing up a provincial constitution without help from educated Eastern Highlanders. Post-Courier, 31 January 1975 and 9 November 1976. See also Good, 'A Note on the Production of the Educated Petty Bourgeoisie' in Amarshi, Good, Mortimer, op. cit.

[97] An interesting development was the resignation of the deputy secretary for Foreign Affairs and Trade, Ben Sabumei, son of the Eastern Highlands' former national parliamentarian Sabumei Kofikai, in March 1978, in order to take up an executive and well paid position with the major coffee exporter, ANGCO, Post-Courier, 30 march 1978.

[98] National Statistics Office, Summary of Statistics 1979, Port Moresby, 1982, Table 67.

6 Coffee consolidation, the development corporations and the rich rural classes

At the end of the decade in which independence came to Papua New Guinea, substantial moves were underway to strengthen rural capitalism in the Eastern Highlands and the leading social classes within this system. National coffee production had increased through the 1970s, especially from 1973 to 1974, but had fallen from 43,819 tonnes in 1974 to 38,064 tonnes in 1976. In line with this, production on the average coffee smallholding dropped considerably from 0.14 tonnes to 0.08 tonnes between 1973 and 1976. In 1979 the national government recognised a serious fall in agricultural export production, brought about, it was officially said, by generally poor husbandry and maintenance, poor management, weak extension and research facilities, a lack of investor confidence, and difficulty in obtaining land for new plantation development. [1] The subsequent strengthening of agricultural capitalism in the Eastern Highlands was, however, a broad, deep and rather original movement borne upon the action of the state and of highlanders themselves. It focussed upon the expansion of coffee production and the consolidation of land under the crop, growth of the development corporations, and the entrenchment of the rich rural classes.

The origins and aims of consolidation

The consolidation of coffee-producing lands in the Eastern Highlands began in 1978 as part of a national programme to boost output in the industry, and was associated with Julius Chan as minister for Primary Industry. The Department of Primary Industry in the province acknowledged the failure of its extension work in recent years, failure which related to short staffing and the inexperience of its extension workers and the reluctance of didimen to go out on foot and offer close supervision to coffee farmers. The consolidation of land by individuals and groups into large commercial ventures, on the other hand, would increase plantings and production in the key and already most developed areas of the province where land was relatively short, and would facilitate the introduction of increased levels of management and technology under conditions which could be adequately supervised by extension workers. [2] But the consolidation of land did not derive from governmental aims and initiatives alone. While consolidation originated with DPI in 1978, it was noteworthy that in the same year the department in Goroka was 'flooded with people' wanting help to consolidate. [3]

The new national production target was set at 60,000 tonnes by 1986 possibly rising to 70,000 tonnes in 1988, an expansion of output which required new plantings, over and above replacement plantings, of some 13,500 hectares. In line with the national policy on coffee development, the Eastern Highlands was allotted 1,300 hectares to be established in 1982. Within the province the allocations to the sub-provinces were as follows. [4]

Goroka	280 hectares
Kainantu	280 hectares
Okapa	220 hectares
Marawaka	200 hectares
Lufa	160 hectares
Henganofi	160 hectares

The increased concentration on coffee production was to be achieved entirely through plantation-type development, on basically two models, one of 15 hectares and over, and one of 5 to 15 hectares.

Certain procedures existed for the establishment of 15 hectare plus plantations. Upon receipt of a request for assistance from a group of farmers, DPI arranged for their registering as a Business Group, the surveying of their land, and the issue of a land title to them. These critical initial procedures were simplified in the case of the individual land owner. The department advised the group or individual on correct farming methods, and compiled a project development programme and submitted an application for a rural loan together with a cash flow budget to the Development Bank on their behalf. [5] The Development Bank required the group or individual owner of the project to provide all of the labour requirements up to the completion of planting, as the project owner's equity, 'sweat equity', in the plantation development. The Bank then provided loan funds to cover maintenance costs and capital requirements through to production, [6] the first harvest occurring in four years.

The vital element of skilled management was provided by the National Plantation Management Agency (NPMA), which supplied an official to oversee the project and to train an assistant manager who was a member of the group or employed by the individual owner. The NPMA was paid for its services out of a management component in the finance provided by the Development Bank. The owner's land title was held by the Bank as its security for the loan. The overall emphasis in the 15 hectare coffee plantation was on efficient management and its costs and on wage labour. The development of this model was, significantly, to be 'actively encouraged' by DPI. [7]

With the smaller 5 to 15 hectare model, no leasehold land title was actually required for security by the Bank, but less material encouragement particularly in the form of bank finance was provided. No loan funds were available for fertilisers, management fees and assistant manager's salary, and the project owner was required to provide a greater amount of equity in the form of labour requirements for the maintenance after planting. Depending on the actual size of the farm, the Bank would finance only 30-40 per cent of such labour costs. Funds were not available for the hiring of NPMA services, and the individual or group owner was dependent on DPI's possibly inadequate advisory services over the life of the project. With the smaller as with the larger plantation, however, all labour was to be purchased. [8]

DPI had specially prepared itself for the implementation of the coffee consolidation programme in the Eastern Highlands. Where the didimen had failed through the mid-1970s a new, small and high powered Coffee Development Team, with a specific task in front of it, might succeed at the end of the decade. In the Eastern Highlands the special team consisted of three rural development officers, one rural development technician, and two rural development assistants. Attached to the group in 1979 were two teams of surveyors provided by the Department of Lands to assist in the surveying of the projects.

Land usage and ownership

Land ownership practices and policies were at the very base of coffee consolidation, and here much greater flexibility and more potentialities existed especially in relation to larger commercial projects than is generally recognised by some observers. [9] The reality of the late 1970s was indicated in Ward's research which showed inter alia [10] 'widespread' individual development of and a 'great deal' of cash crop production on customary land; that individual enterprises were 'most numerous' and there was 'considerable evidence' that groups of Papua New Guineans were producing cash crops on customary land and 'modifiying traditional tenure rules for the purpose'; that 'considerable evidence' existed of a demand for official formalisation of agreements and sales, and there was 'widespread popular support' for direct land dealing, as 'straight commercial transactions [were being] elaborated under the guise of "customary" dealings'; and that tendencies existed towards 'high land prices and exploitative landlordism', for example, 'soaring' prices in the Wahgi Valley as 'several thousand' migrants bought land from local clans. The reality of the period was also alluded to in an information booklet of the Land Court Secretariat which, in the first paragraph of the first page, declared directly: 'Many people in Papua New Guinea own land ... They earn a living by growing crops, raising animals or selling the products of their land. Land is very important to these people'. [11]

The fact that freehold remained an uncommon arrangement of land ownership among Papua New Guineans was neither the most significant aspect of the problem nor an insuperable barrier to commercial development. [12] Land held under customary tenure could indeed be mobilised for larger business proposals, as another official publication helpfully explained. [13] While 'the major limitation on using land under customary tenure [was] the difficult legal status of the land and the owners', a 'useful service' could be provided under existing legislation.

> The State can provide a title to the extent that it holds an estate in land. It can lease land if it holds a leasehold interest in land ... Such a sublease provides a fairly satisfactory title over the land to secure a development loan. Such a sublease may be to an incorporated group of essentially the same owners who lease the land to the Government. This then becomes a lease and lease back arrangement. [14]

It was pointed out that, under this arrangement, a lease without right of renewal could provide sufficient tenure for the production of tree crops such as coffee and cocoa where the project was serviced by an existing marketing or processing organisation and accommodated on land secured for a single crop cycle of some twenty or thirty years. [15]

What could be done by the individual or group land owner was not necessarily what would be done, however, in terms of the extension of all-important government resources for commecial development. The cost of all the processes of leasing the land and leasing back were significant, and 'preference should be given to larger projects'. The requirements of an attractive return to clan-group members suggested to the state that 'projects requiring a development loan exceeding K25,000 should be acceptable and smaller projects might be handled if staff [were] not fully committed'. But loans that did not significantly exceed K10,000 could often be handled by customary land agreement, and 'it [was] not efficient use of staff to provide a lease back arrangement for these'. [16] As already indicated, this preference for assisting the large commercial project, of giving to those who already had, was perfectly consistent with the rural development strategy established over almost three decades by DASF-DPI and subsequently by the Development Bank. [17]

In the case of the 5 to 15 hectare plantation model, the legal status of the land, as required by the Development Bank and fostered by DPI, was simply the registered Clan Land Agreement. If disputes existed among the so-called traditional owners of the land in question these tended to appear and to be dealt with when land surveyors started work. [18] When official mediation was necessary in any disputes of this kind, for purposes of consolidation or otherwise, the resulting agreement could be approved by a Land Court and became an order of the court. [19] The registration of a clan land agreement in similar fashion strengthened this ownership arrangement for the people concerned. Land held under this simple and limited form of ownership was, as noted, capable of attracting loan finance and of being put to work commercially.

The legal status of the larger plantation model was in 1979 usually that of the twenty year Special Purposes Lease. The leasehold helped to obtain for the owners of these plantations a relatively high level of bank finance, but it was nevertheless a rather loose and inadequate arrangement in the view of DPI. The government leased only the land itself to the project owners, and did not include the value of the assets, such as coffee trees, on the land. The land supposedly reverted back to the original owners, at least some of whom might differ from the lease and project owners, on the expiry of the lease, but the mechanics of this process had not been worked out in 1979. The Development Bank had insisted on a leasehold arrangement for the large plantations so that it could claim the land back if the project failed, but it was to be doubted that the Bank would in fact succeed in such action especially if directed against original owners among the leaseholders of the consolidated blocks. Such limitations prompted Murray of DPI to press for freehold rights for the owners of the large plantations, a move which naturally was strongly favoured by the owners themselves.

DPI officials were also concerned with some associated limitations on the profit making potentialities of the larger blocks. Bank loan interest rates, 9.5 per cent per annum in 1979, and management fees were high, and the owners had to wait five years for a return from their coffee. The position of the consolidated block holders was particularly bad, in the view of Nitsche and Murray, in comparison with coffee farmers who acquired property under the older Plantation Acquisition

Scheme operating simultaneously in the country at the time. Here a group raised only a small part, some 20-30 per cent, of the agreed value of the plantation as their initial down payment, and the balance was paid by the government to the outgoing, previously always foreign owners as an interest free loan to the local group. Plantations acquired under this scheme also provided their new owners with an immediate return from coffee. It was clearly a much more financially attractive scheme than the consolidation programme. Actually existing state policy therefore favoured takeovers, which moreover were no longer being directed solely against the foreign plantation owner, a rapidly declining breed in the Eastern Highlands at the end of the 1970s, but against such Papua New Guinean owned organisations as the coffee marketing group ANGCO and the development corporations which were such a distinguishing feature of rural capitalism in the province. The indications were that such takeovers would grow. Plantation takeovers did not boost coffee production at all, and Nitsche and Murray therefore favoured tightening the conditions associated with such acquisitions while relaxing those for consolidated block owners. The national government announced a so-called freeze on plantation takeovers, in August 1979, until problems in the scheme were ironed out, but exempted transactions not then finalised and strictly private dealings. [21]

The bringing of new land under coffee in a region already extensively given over to the commodity had inherent problems. It was estimated in the early 1970s that some 77 per cent of the total area of the Eastern Highlands was unsuitable for agriculture, that only 1.2 per cent or some 13,000 hectares were in fact suitable for all crops, and that the Eastern Highlands already had 'a cultural if not an absolute land shortage'. By 1976 coffee was being grown on some 11,760 hectares of the best land in the province. [22] The land on which consolidated blocks were being established in 1979 was mostly steep or gully ground previously unused for many years. Murray estimated that some three-quarters of the land marked for consolidation at the time was suitable for coffee and that the rest was probably too steep. [23]

It was clear that in the Eastern Highlands at the end of the 1970s land was being actively and enthusiastically exploited under a variety of arrangements by both groups and individual Papua New Guineans - under registered clan agreements, various forms of leasehold, freehold and, of a certainty, various direct but informal agreements between sellers and buyers. The acquisition of existing coffee plantations had made considerable progress in the Eastern Highlands and in the Western Highlands, the two largest coffee producing provinces in the country. At the end of 1977 only some 38 per cent of Eastern Highlands plantations were under indigenous ownership, while approximately 62 per cent were still foreign owned. In 1979, however, most coffee plantations in the province were owned by Papua New Guineans, in favourable contrast with the national situation for all types of plantations where more than half remained in foreign hands. The following table is indicative of the progress in the highlands by early 1979. By the end of the year further transactions had occurred and 16 plantations in the Eastern Highlands had been brought under the control of NPMA.

Table 6.1

Change in ownership of coffee plantations, Eastern, Western and Southern Highlands, June 1979

Eastern Highlands Province

Plantation	Date of takeover	Acreage	Current owner/manager
Kami	27.1.78	93	N.P.M.A.
Kiloparoka	17.5.78	95	N.P.M.A.
Upegu	2.3.79	135	N.P.M.A.
Aromet	1.9.78	120	N.P.M.A.
Huvei	13.3.79	232	N.P.M.A.
Lahamenegu	1.7.77	50	Gouna Development Corporation
Erinvale	1.7.77	170	Gouna Development Corporation
Roka	1.4.78	110	Gouna Development Corporation
Aionora	May 1977	130	KKB Limited
Kawaita	1971	110	KKB Limited
Tuta	1979	80	KKB Limited
Numbia	May 1978	160	Highlands Henene Pty Ltd
Karangka	August 1978	80	Highlands Henene Pty Ltd
Hobe (including Muriho)	May 1976	100	Benbosi Pty Ltd
Asaro Coffee Estate		170	Asaro Coffee Estates
Obihaka		140	Asaro Coffee Estates
Clearence		250	Asaro Coffee Estates
Foinda	1974	80	Asaro/Watabung Rural Development Corporation
Kesa	1974	75	Mr. Bono Azanifa
Mountain Dale (Anego)	18.7.75	125	Mr. Akapite Wamiri
Bena	1975	178	E.H. Development Corporation
Sogoperu	1975	220	E.H. Development Corporation
Matus	1977	120	E.H. Development Corporation
Kapakamaliki	1975	178	E.H. Development Corporation
Keya	1975	71	E.H. Development Corporation
Duna	1977	80	Duna Coffee Estates
Fano	1977	80	Duna Coffee Estates
Kompafo	1977	70	Duna Coffee Estates
Korona		200	Mr. J.W. Wan
Kamalaki		60	Catholic Mission
Erimissoo		95	Mr. W. O'Brian
Tapo		65	Australian Overseas Development Corporation
Akwitanu		82	Bena Coffee Lands
Nori Kori		247	Bena Coffee Lands
Ambaura		70	Bena Coffee Lands
Bonta		70	Australian Overseas Development Corporation
Arau		106	Mrs. M. Larner
Abiera		30	Development Bank

Plantation	Date of takeover	Acreage	Current owner/manager
Aiyamontina		40	Owned by a National
Kama		77	Mr. G. Bluff Brugh
Urara		200	Kainantu High School
Tudor		30	Uncertain
Arara		110	Bena Coffee Lands
Korfena		160	Steamships Trading Company
Gisumonoso		82	Steamships Trading Company
Wantorifu		70	Steamships Trading Company
Kenimaro		110	Steamships Trading Company
Western Highlands Province			
Waraka	1.4.74	133	Wahgi Mek Pty Ltd
Jimbina	25.8.75	160	Wahgi Mek Pty Ltd
Sigri	1.4.74	75	Wahgi Mek Pty Ltd
Banz	1.4.74	40	Wahgi Mek Pty Ltd
Talu	1.4.76	65	Wahgi Mek Pty Ltd
Mintal	°1.4.76	170	Wahgi Mek Pty Ltd
Karamang	1.4.76	76	Wahgi Mek Pty Ltd
Wahgi	1.11.78	250	Wahgi Mek Pty Ltd
Mangara	1.11.78	400	Wahgi Mek Pty Ltd
Darminie West	1.4.77	10	Wahgi Mek Pty Ltd
Nondugl	1.4.78	80	Wahgi Mek Pty Ltd
Kul	23.2.77	158	Wahgi Mek Pty Ltd
Amuliba	17.4.78	212	N.P.M.A.
Warawagi	19.5.78	160	N.P.M.A.
Kalanga	19.5.78	53	N.P.M.A.
Kidjibi	19.5.78	280	N.P.M.A.
Bitam	1.9.78	135	N.P.M.A.
Kimel	18.6.79	350	N.P.M.A.
Tremeane	18.6.79	300	N.P.M.A.
Wurup	May 1975	220	Pipilka Development Corporation
Tigi	November 1976	210	Dei Council
Gumanch	November 1978	1,000	Coffee International Pty Ltd
Koban	19.6.78	400	Harcros Plantations
Karmal	19.6.78	200	Harcros Plantations
Kuge		200	Mr. J.E. Leahy
Korgua		80	Mr. D.J. Leahy
Melda		140	Mr. C. Leahy
Pabarabuk		90	Mission Owned
Komun		100	Dr. D. Manton
Kigabah		260	Investment Corporation (ANGCO)
Baglega		180	Kagamuga Trading Co. (ANGCO)
Wagil Sipia		225	Karu Corporation Pty Ltd (ANGCO)
Koibugga		80	Mr. L.P. Bliaux
Alimp North		130	Verona Coffee Estates (ANGCO)

Plantation	Date of takeover	Acreage	Current owner/manager
Southern Highlands Province			
Alimp South		--	----
Koropingi		40	S.H. Development Authority
Lamane		30	S.H. Development Authority
Mesebem		30	S.H. Development Authority
Turama		20	S.H. Development Authority
Kamp		30	S.H. Development Authority
Hala		38	S.H. Development Authority
Turupu		30	S.H. Development Authority
Okuafugua		30	S.H. Development Authority
Alipugu		30	S.H. Development Authority
Obia Tangi		30	S.H. Development Authority
Batame		30	S.H. Development Authority
Mungku		30	S.H. Development Authority
Tiri		30	S.H. Development Authority
Pondiabo		30	S.H. Development Authority
Erave		116	Mr. R. Neville

Note: The Southern Highlands Development Authority managed all the province's plantations on behalf of the various Land and Business Groups.

Source: Coffee Industry Board, Goroka.

The applicants for consolidation

In late 1979 DPI in Goroka had received some 100 applications for consolidation, and sixteen projects had been approved by the Development Bank. [25] Between ten and twenty applications were under consideration by the Bank, which acted slowly and bureaucratically in Murray's view. About ten projects were already under coffee and by the end of the year thirty projects were expected to be in operation. Most projects were located in Asaro and Bena near Goroka, with some also in Okapa and a few in Kainantu, where more were expected in 1980. There were a few consolidated blocks of 30-50 hectares, mostly in Bena, but the average size in November 1979 was some twelve hectares.

The case of an applicant in November from near Kainantu was possibly indicative of some of the operating norms within individually-owned projects. He was a confident man of approximately thirty years who came from the Kainantu area but had leasehold ownership of 145 acres further out, that is, land over which he had no clan affiliation. The land originally had been intended for cattle, but its then owner had defaulted on a Development Bank loan and the applicant had taken over the lease. He planned to employ as assistant manager a man from his own village but of a different clan, and to pay him K40 per fortnight. It would be the assistant's task to actually manage the coffee project while the owner, who would not live on the plantation, was busy buying coffee with his four vehicles. The project owner was able to contribute K5,000 to the project immediately, for building a house for the assistant manager, for clearing and fencing, and for buying some 20,000 seedlings, some of which would be got from the Kainantu High School, and some in classic 1950s-1960s development style, from his friend and former employer, the planter Neil Lambert. The plantation boundary was already cleared and fencing, with a creek on one side, was estimated to cost K1,370. The owner planned to employ ten labourers and to pay them K15 each per fortnight, a very low but not untypical wage rate. He had received previous loans from both the Development Bank and the Banking Corporation for his vehicles and for commercial vegetable growing, and he repaid them in full despite the failure of the vegetable project. The applicant wanted the Development Bank to pay his wages bill in the consolidation project. [26]

Of the applications received by DPI in 1979, approximately twenty per cent were individually owned projects and were generally smaller sized. Consolidated blocks were individually owned by Sailas Atopare (15 hectares) and by Gai Duwabane (10 hectares) and another was being proposed by Paul Asaro. The first two were important national politicians and the last named an official of both ANGCO and the Bena (renamed the Eastern Highlands) Development Corporation, and all three representatives of the rising rich rural classes, the rich peasantry and rural capitalists, active in rural capitalism in the province. [27].

Table 6.2
Consolidation blocks, Eastern Highlands, 1979

Name	Location	Area (ha.)	Amount of Loan (K)
Holipoha	Kafogu	56.7	35,000
Gizowa	Ahamehu	14.4	10,000
Lusaro	Kerefa	9.0	2,000
M. Korosahu	Asaroha	30.0	24,100
S. Atopare	Kabiufa	15.0	10,000
Nipijaleha	Masilokaufa	47.0	20,000
Warigiso	Kafana	19.0	10,542
Kahae-Minoho	Kafana	11.0	5,000
G. Duwabane	Kofena	10.0	5,222
Sonohini-Raharupi	Korofaneta	23.0	-
B.G. Kerefa	Kerefa	8.0	-
P. Asaro	-	-	-

Source: DPI, Goroka.

Group owned projects were formally composed as a Business Group, often of two or three families, with a management committee supposedly empowered to determine the allocation and distribution of eventual proceeds. But it was noticeable that groups almost always had one or two individuals in the forefront. Located outside Goroka on the road to the Kotuni trout farm, the Gahuku village group, for example, had a planned project of 18 hectares of coffee including perhaps 30 per cent of already established coffee. Harry Gotoha had encouraged the owners of five coffee farms to give up their individual rights to the consolidated group. Gotoha was in fact the 'major force', in Murray's view, within the Gahuku plantation group, and was perhaps the richest Papua New Guinean rural capitalist in the province. Other potent commercial and political figures active in groups involved in the consolidation program were Auwo Ketauwo and Isembo Olekavea, and Bebes Korowaro, a lesser figure yet quite unlike the average highlands coffee grower, was also prominent in consolidation. [28]

The Eastern Highlands was well advanced on the consolidation programme in relation to the other coffee producing provinces in 1979. Land disputes were delaying consolidation in the Western Highlands, the Southern Highlands had a different approach to coffee production, and Simbu was seen by DPI as having a low potential for the 15 hectare scheme, with the possible exception of Kerowagi. Simbu was moreover accorded a special role in furthering consolidation in the Eastern Highlands in the planning of DPI.

Wage labour, management, finance and innovation

The Development Bank and DPI calculated that, for the establishment of a twenty hectare consolidated plantation, 2700 days labour per annum were required. This represented a requirement of fifteen full time workers in the first year of the project. The second year required 2,900 days or the full time work of twelve, a need that was more or less constant until the end of the fifth year. In the sixth year the labour requirement fell to 1,600 days, but it rose again in the seventh year to 2,200 days and remained constant at this figure thereafter. Harvesting from the fourth year would call for additional labour for short periods. [30] Anderson's 1976 survey of coffee producers had shown that the ability to mobilise labour was of 'utmost importance' in the efficiency and good management of a holding. [31]

Much of the initial labour for clearing and for the establishment of the consolidated blocks was expected to come from local villages. The labour requirements of the new plantations, however, were expected to be fairly significant. If the 1979 average size of twelve hectares per block was maintained, the Eastern Highlands would establish some 190 projects, DPI calculated, by the mid-1980s. [32] Official planning emphasised that after production began all labour for consolidated blocks would be on the basis of fully paid wage labour. Local villages in the Eastern Highlands were not alone expected to meet the demand, and many future labourers were to come from Simbu. [33] This province's low potentiality for consolidation would work in favour of the development of consolidation in the adjacent Eastern Highlands, even more firmly relegating Simbu to the position, which it had occupied since the 1940s, of a labour reserve for the Eastern Highlands and the other advanced regional agricultural economies. In the process Simbu would fall behind even further in rural capitalist development.

Anderson's survey stressed the importance of good management, and had seen it related to both the size of the holding and to the use of wage labour. [34] Efficient management would now be built in to the functioning of the consolidated blocks through the services of NPMA. The state agency provided one manager to oversee the operations of six plantations and to train an assistant manager who was a member of the group or employed by the individual project owner. [35] DPI had tried to get the Gouna Development Corporation to act as a management agency, but Gouna had declined. The Bena Development Corporation and ANGCO, the latter being the management agency for Bena itself, were interested in the proposal, but ANGCO said that the consolidated blocks would have to market their coffee through them, and DPI saw this as unreasonable. Many Eastern Highlanders involved in the programme, some of whom were among the biggest producers in the province, had wanted to provide management themselves, not least because of the costs of the NPMA's services. Murray of DPI wished to see NPMA's services cutting out in the fifth year of the project, when the individual or group had experienced two harvests, and the assistant manager had seemingly acquired the ability and earned the right to run the plantation himself. [36]

The Plantation Management Training Programme of the Department of Commerce was another answer to the problems of inefficient indigenously owned plantations. A nine month programme in coffee plantation management was introduced in late 1979, and it was estimated that the highlands required some 200 managers over the next four years. The first group of trainee managers totalled fifteen of whom the Coffee Industry Board sponsored eight, plantation owners and employers five, and trainees' families the rest. Professionalism was strongly emphasised. Project officer Robert Henderson said, 'Plantations are businesses and we aim to have them run as such. A lot have gone bust because they have not been run as businesses but only as status symbols'. Trainees also appeared to have a commercial and realistic approach. Elijah Giheno was from the Eastern Highlands where his family were apparently in the process of acquiring a 50 acre plantation about 30 miles from Goroka. Although he came from an area where coffee had long been grown, Giheno recognised that he 'did not know very much about it' before his course began. 'Walk round the highlands and you see that coffee really needs attention', he added. The problem itself and the pressures for the establishment of the training program from national and provincial governments, the coffee industry, and individuals ensured employment to all qualified managers. It was a seller's market, Henderson observed. [37]

The total loan that was available on the basis of the Development Bank's larger model was K69,070. Repayments were estimated to total K124,929 (at the interest rate of 9.5 per cent), and to be completed in the fourteenth year of the project. As things stood, the group or individual project owner was expected to obtain a surplus after expenses of at least some K20,000 in the seventh year of the plantation's development when full production was begun. [38] The income earning capacities of the plantations were thus in no way negligible.

Management fees were, however, a large part of the development costs of the plantation. They would total K5,000 in the first year of the project alone, and while Village Economic Development Funds were supposedly available to new block owners, these funds were believed to be inadequate by DPI in 1979. [39] DPI's attitude to the funding of consolidation emphasised not the acquisition by a fortunate group or individual of a valuable and profitable plantation with the vital assistance of the state, but typically the supposedly burdensome nature of certain elements of the programme such as the financing of the management services. Murray noted that if the management fee was removed after the fifth year of the project the group or individual owner would enjoy a reduction in costs of K50,000 by the fourteenth year. He and Nitsche were impressed by the innovative quality and productive potentialities of consolidation both in itself and by comparison with the less dynamic and increasingly inward turned and non-productive Plantation Acquisition Scheme. The DPI officials felt that the government should subsidise loans to consolidators and reduce their interest rates, while desirably, substantially increasing interest rates on plantation take- overs to 12-13 per cent. They related these proposals to their above noted preference for the granting of freehold rights to consolidated block owners. In ways similar to those of McKillop and other DASF/DPI officials during the early expansion of indigenous coffee production, Murray and Nitsche identified closely with the innovative, entrepreneurial and large scale coffee consolidators, and they pressed for better government assistance to them in order, as Murray put it, 'to give these guys a better deal'. [40]

Large scale land ownership, specialised extension services, skilled management, wage labour and bank finance, represented in combination a greatly improved technical basis for indigenous coffee production in Papua New Guinea. Commercial coffee growing had been established in Papua New Guinea with the Blue Mountain variety of arabica, but experiments at the highlands agricultural experimental station at Aiyura showed that two varieties, Bourbon and Arusha, both outyielded Blue Mountain by 'a considerable margin'. After experimenting with spacing, pruning, shade control, and fertilisers, Aiyura recommended the two new varieties, and began distribution of the seeds through the 1970s. [41] Aiyura also experimented with high density plantings, up to 10,000 trees per hectare. It found that high densities were much more productive than lower densities in the early years of a tree's life, but unless good pruning management was applied they became relatively unproductive in later years because of canopy closure. If the much more productive high densities were to be properly utilised, a strong emphasis on pruning technique was demanded. [42] Improved varieties and techniques were thus available to the new consolidated block owners, whose skilled management resources were better able to meet the necessary pruning requirements than was the small scale owner. The consolidated plantations, with their trained management and their permanent wage labour were also well capable of utilising new machinery when that became available.

The growth of the development corporations

The Goroka local government council formed its development corporation or 'business arm', Linupa, in the mid-1970s, but the corporation became really active in the economic life of the province only near the end of that decade. [43] It purchased a service station in Goroka, and dallied with a number of car import dealerships before settling with Mitsubishi/Chrysler trucks and cars, and Honda motorbikes, through the Toba company of Port Moresby. It similarly acquired an insurance agency (CIC) and a car rental agency (Budget Rent-a-Car), the latter on the basis of Linupa paying for the agency name while owning the cars. By 1979 its largest activities were in hotel ownership and management. Goroka's Minogere Lodge which the council already owned was expanded through the addition of a large disco-bar, and the resulting transformation was heralded by the new name of 'Linupa Minogere Lodge' and the introduction of higher tariffs. More significant, however, was Linupa's purchase of the Lantern Lodge, in many ways Goroka's premier hotel, in July 1979. The local government development corporation made the acquisition by borrowing K300,000 from the new Goroka branch of the Australia and New Zealand Bank (ANZ). Godfrey Benjamin, Linupa's business manager, noted that the ANZ branch was very keen to get the group's account, and the method of purchase was indicative of the strong fund raising capacities of the government development corporations. The services of the hotel's well known Chinese chef were retained, while Linupa extensively renovated the establishment's Brick Bar, and a turnover of K15,000 was obtained over the 1979 Independence weekend alone.

An important characteristic of all four Goroka based development corporations was their keen awareness of the value of good management services and training, and with Linupa it was San Miguel and Swan Holdings, the brewing companies, which provided management services and staff training in 1979. Benjamin felt in September that Linupa had 'grown too fast' in the two previous years, and he aimed to consolidate control over the properties already acquired. Problems existed with the loss of newly trained staff, and a new programme of training and promotion had been introduced with the help of San Miguel.

Characteristic also was Linupa's close relations with the other provincial development corporations. This was especially so with the provincial government's Nokondi. Two of Linupa's directors, including Benjamin, were represented on Nokondi's board of four directors, and Benjamin was not asked to step down when he resigned from the provincial government to join Linupa. Both groups were involved in collaborative fashion in car rentals in Goroka. Good relations also supposedly existed with the Gouna Development Corporation. Linupa owned the British Petroleum agency in Goroka before Gouna purchased the Mobil agency in 1978, but neither group intended to steal the other's customers according to Linupa. What was important in these good relations, said Benjamin, was the personalities of the directors concerned. The province's national politicians were specifically excluded from this happy interrelationship, however, being only 'out for themselves', and uninterested in the administrative affairs of Linupa. [44]

Nokondi was also very active in commercial, mainly urban, property ownership. [45] It retained holdings in 1979 in Gouna Development Corporation, Pacific Helicopters, and Kotuni trout farm, and in a number of Gorokan commercial ventures. It owned Kwan's garage in the town, and held title over vacant and valuable land in West Goroka. It had two subsidiaries and planned to enter into joint ventures with established companies in hardware and in plastics manufacture. Rather like Linupa, as a government corporation it occupied a superior borrowing and financial position. This was seen when Nokondi acquired James Arcade in Port Moresby, with the assistance of a K100,000 interest free loan from the national government. The Port Moresby property had proved very profitable, and Nokondi intended to resell it to an existing willing buyer when the time was right. Nokondi's commercial attractiveness and advantages were subsequently demonstrated when it obtained 17 per cent of Collins and Leahy Ltd of Goroka, at a cost of K700,000, on the basis of full bank finance from the Papua New Guinean Banking Corporation. Collins and Leahy, in the view of provincial secretariat Executive Officer McConaghy, was anxious to obtain some 'political back up', and Nokondi was able and willing to provide it. [46]

Nokondi displayed a preference for urban commercial property ownership, but it considered possible future coffee production. It had an interest in land totalling 1,400 hectares at Yonki near the Ramu hydroelectric station, and Nokondi was seeking to lease the land for coffee, in conjunction perhaps with ANGCO as management. Nokondi was aware that ANGCO and other companies were keen to obtain investment funds from the provincial government corporation.

The number and strength of development corporations in the Eastern Highlands had grown by late 1979, and McConaghy saw the possibility of 'incestuous relations' growing between them. The provincial government therefore considered the establishment of a coordinating agency to oversee their activities. Nokondi's position vis-a-vis other corporations and companies in the province was based on financial and commercial preferences and on political-legal superiority. Only the provincial government had the capacity to make regulating arrangements for the development corporations and to extend support to Collins and Leahy. The other corporations simply emphasised the existing policy of good relations between the owners and directors themselves.

Gouna Development Corporation, one of the two leading non-government, quasi-populist development corporations of the Eastern Highlands, was still operating in terms of its earlier diversification plans, and by September 1979 its assets totalled K1 million. [47] It was expanding its poultry, intending to reach some 20,000 hens in early 1980, with arrangements whereby day old chicks came in by air from Brisbane, and feed from Lae. Gouna's poultry manager was a New Zealander and the assistant manager was a well educated local man. The corporation was considering closing the piggery, however, due to shortage of space at Lahamenegu plantation where there were poultry, pigs and coffee. This (valued at K400,000) was one of the three plantations then owned by Gouna, the others being Erinvale (value K167,900), and Roka (value K396,200). Sixty acres of new coffee on new land had been planted at Erinvale, and Gouna felt that coffee was going well, providing extra cash flow.

The policy of diversification prompted Gouna to acquire the Mobil agency, involving a depot and freight in the province and refuelling at Goroka airport, in late 1978. It opened a new tyre and battery service at North Goroka in August 1979, moving its existing company (Goroka Tyre and Battery) to the new site, and meeting the costs of expansion, K50-60,000, from its own assets. New machinery was installed at its Goroka Soft Drink Company factory, where Gouna found that it could not produce enough of its 'Laga' soft drinks for the booming all-highlands market. Gouna considered entering car sales in future, but it basically planned to improve and extend existing businesses. Pacific Helicopters, for example, was a very profitable company, its helicopter fleet was rapidly growing, and indigenous ownership was fast expanding also. [48] Gouna remained apparently uninterested, on the other hand, in gaining ownership of Goroka's Bird of Paradise hotel and, as Berqvist said, would not 'go in for everything'.

The development corporation faced trouble, however, on two fronts. Roka plantation, acquired in April 1978, was subject to a takeover claim by a group of local landowners. Roka's former absentee foreign owner had wanted to sell for four years before Gouna obtained leasehold ownership, said Berqvist, and local landowners had remained silent. The claimants were now being led by national politician Gai Duwabane (himself a Gouna member) and his mother. The development corporation felt that the action was not truly representative of local feelings, and it was concerned about the claim's broad significance whereby the national government could give a group of people the right and money to take an efficiently run plantation from an Eastern Highlands-owned development corporation. A delegation of Gouna directors had taken the matter to Prime Minister Somare in Port Moresby.

Gouna was also still concerned about the role of Nokondi and Linupa. 'Unfair competition' was involved, said Berqvist, in the ease with which the government corporations obtained bank finance and loan money. Gouna paid taxes which effectively assisted the government corporations, while the Goroka local government council, for one, seemed more interested in furthering the business activities of Linupa than in its proper tasks of road mending and so on. But the links between Nokondi in particular and Gouna were strong. Gouna retained sizable investment funds from Nokondi, a Nokondi director was on the Gouna board (after the resignation of Paul Ine from both Nokondi and Gouna), and Gouna directors Ketauwo and Olekavea were, after 1978, provincial government members. Nevetheless Berqvist believed that it 'could come to the stage of bad feelings' between Gouna, on one side, and Nokondi and Linupa, on the other.

Bena Development Corporation in 1979 remained strongly committed to rural, not urban, development and to coffee production particularly. [49] Its ownership had grown to nine plantations, including five at Bena (Sogopegu, Bena, Magitu, Kapakamaliki and Keya), three at Duna

(Duna, Kompafo and Fano), and one at Kainantu (Matus). The acquisition of the Kainantu plantation, in late 1978, further illuminated the essential flexibility of ethnicity and of clan land owning arrangements, as did the fact that Akepa Miakwe and Paul Asaro were inviting villagers in the vicinity of Matus to become shareholders in Bena, with Miakwe travelling to Kainantu to attend to this personally. The corporation planned to establish cattle at Matus to supplement the existing coffee. Similarly to Gouna, Bena's Duna plantation was, however, subject to a takeover bid, with educated people, in Miakwe's view, promoting the agitation among local villagers. A Bena delegation had visited Somare and Lands minister Boyamo Sali in Port Moresby, and a letter was sent to the prime minister from Gorokan businessmen. With almost all of its resources committed to commecial coffee production, Bena was seemingly under a greater potential threat from the functionings of the Plantation Acquisition Scheme than was Gouna.

Bena maintained good relations with Gouna, and had no problems with Nokondi, in Asaro's opinion, but Miakwe still preferred provincial and national governments to stay out of business. This relatively low level of concern was presumably related to the fact that Linupa and Nokondi's operations were mainly in the urban and commercial sectors. Bena too believed in the importance of skilled management, and while it aimed at getting indigenous managers on all of its plantations in future, it intended to renew its management contract with ANGCO for a further five years when the existing contract expired in 1980.

Food production

DPI in Goroka admitted to a 'bit of a worry' from the impact of coffee consolidation on already weak food production in the province. Food and nutrition were supposedly the first priority of DPI in the province in 1979, but actual plans were modest. Phase one of a five year programme for commercial vegetable production in 1980 called for the establishment of only seven projects of some five hectares, each built upon existing provincial imbalances, Goroka would get three projects, Kainantu two, and Lufa and Henganofi one each. [50] None of this was new. The cold reality was that 'not much money was to be made in vegies', as Murray and a failed vegetable grower and applicant for consolidation both agreed. [51] Gouna corporation, however, had become a successful producer of poultry, which indicated what might be done in particular areas of food production on an appropriate scale and with adequate financial resources and management. In a province where almost all good land was already established with coffee, consolidation would take much of what little remained away from the growing of food. The impact of coffee on food production had been suggested in a number of studies. Salisbury said in the 1960s that coffee took the best land with the shortest fallow period out of the subsistence crop rotation cycle, causing the average fallow period to increase and requiring in consequence much more land to produce food. Harris estimated more recently that while a sweet potato grower needed only 0.07 hectares of best quality land to produce all food requirements allowing for periods of fallow, the amount of land required to grow sufficient coffee to be able to live on imported tinned fish and rice, the preferred modern alternative to the traditional staple, was 'consistently greater'. [52]

Consolidation would also take labour, and a still greater burden would presumably fall on women as producers of food in the province. Decreasing domestic production and the consequent increasing importation of food were joint processes which would of course bear most heavily on the poor peasant majority in the Eastern Highlands. DPI might at last conceive of the possibility of inter-cropping certain food crops with coffee (specifically mentioned were spices and chillies), but this was unlikely to be more than a palliative given the weight of the problem posed by consolidation.

The strengthening of the rich rural classes

Consolidation represented a renewed, increased, and reactivated commitment by the state to large scale coffee production, and to the rich rural classes already established in coffee. Many of the richest farmers of the province and those striving manfully to emulate their achievement, Gotoha, Ketauwo, Olekavea, Atopare, Duwabane, Korowaro, Asaro, were quickly involved in the rush of applicants for consolidation. As individual land and project owners or as members at the forefront of village and business groups, they were experienced in dealing with the state, in leading villagers, and in the development of organisations. The independence and resourcefulness of these 'influential people' who, for example, readily obtained advice on planting and pruning practice that was contrary to DPI policy, was causing problems for DPI as early as September 1979. [53] But increased state support for plantations had a capacity for exacerbating tensions and rivalries within the rich rural classes, between the established rural rich like Gotoha, Ketauwo, Olekavea, and Akepa Miakwe, on the one hand, and the younger rising rich like Duwabane and Atopare, on the other. Land takeovers led or organised by 'educated men' such as Duwabane were already directed against the holdings of Gotoha's Gouna and Miakwe's Bena corporation, and conflicts over land ownership in the Eastern Highlands were not expected to decrease. But the common interests of the rich classes were expressed and acted upon in the inter-linking directorships of the development corporations and the expressed sympathies of their leaders. The harmonious division of interests between Gouna and Bena, with the latter committed almost entirely to coffee and the former highly diversified, was a clear formulation of how well the rural capitalists of the Eastern Highlands could coordinate their action.

The development corporations were richer at the end of the decade, Gouna's leaders, Ketauwo and Olekavea, were established in provincial politics, and both Nokondi and Linupa were much more economically active especially in urban commerce. The increased intervention of the government corporations into 'business' and the extensive financial and political resources they commanded, was a further source of tension among the rural rich. Through its engagement or otherwise with a particular organisation or activity, or through the exercise of regulatory powers, Nokondi had a particularly strong capacity to affect the interests of groups like Gouna and Bena and disrupt relations existing between them. Other tensions, perhaps of a less serious kind, existed between national politicians, specifically Atopare, Holloway and Duwabane, and the others fully committed to the Eastern Highlands. Miakwe, an older, established and socially well integrated figure, appeared to handle this distinction relatively well.

Consolidation meant also that the highlands of Papua New Guinea would become still more reliant on the fortunes of a particular commodity, with the many consequences therefrom. World agricultural prices are the most unstable of all commodities, and coffee prices fluctuated, for example, between K906 per tonne and K1,931 per tonne between July 1975 and June 1976. [54] Papua New Guinea had developed an ability to meet some of these problems. The Coffee Industry Board's stabilisation fund had, by 1979, accumulated a large reserve, and planned to build storage sheds in Goroka and Lae where coffee could be held when prices fell. World quotas for national coffee production were not then expected, and even if they were introduced DPI felt they would be lifted again before the consolidated plantations entered production. [55]

Consolidation promised to increase the long existing uneven development of agricultural production in Papua New Guinea. The Eastern Highlands, where commercial coffee production began, was off to a quick start in consolidation, and existing plans referred to the subordination of Simbu as labour supplier to the plantations of the Eastern Highlands. Within the province, consolidation was largely concentrated in the vicinities of Goroka and Kainantu, where administrative and economic resources had long been focused, and where, in Goroka in particular, many of the rural rich were of course located. Consolidation involved the state's commitment to the plantation system, [56] so redolent of colonial modes, and intensified plantation production would bring its own characteristic problems of heavy labour exploitation. It would greatly widen the gap between the vast majority of coffee farmers dependent on some one-tenth of a hectare and but a few hundred trees, unassisted by DPI and the Development Bank, and the very few who owned or controlled the large, efficient, state aided plantations with 10,000 trees to the hectare.

Notes

[1] Bureau of Statistics, Summary of Statistics 1975-76, Port Moresby, Rural Industries 1975-76, pp. 5, 51; The National Public Expenditure Plan 1980-1983, National Planning Office, Waigani, November 1970, p. 29.

[2] Personal interview with H.H. Nitsche and Kevin Murray, DPI, Goroka, 20 September 1979 and 'Information Paper for the Members of the Provincial Government and the Cabinet on the Coffee Production Policy for the Eastern Highlands Province', no author, 1979, p. 3.

[3] Interview with Nitsche and Murray, 20 September 1979.

[4] 'Information Paper ...', pp. 1-2.

[5] Included in the loan proposal were provisions for funds for fencing materials, fertilisers, tools and equipment, transport, management fees, assistant manager's salary, workers' compensation, capital requirements, a percentage of the labour costs and contingencies. 'Information Paper ...', p. 3.

[6] Loan funds were actually released only when the block was completely planted.

[7] 'Information Paper ...', p. 3. The National Plantation Management Agency Ltd., a wholly government owned company, was incorporated in February 1977 to act on a consultancy basis for the Department of Lands and to provide management expertise for plantations acquired under the plantation acquisition (or redistribution) scheme which had been introduced in 1974. Maxine Dennis' Appendix to 'Plantations', in Donald Denoon and Catherine Snowden (eds) A Time to Plant and a Time to Uproot: a History of Agriculture in Papua New Guinea, Institute of PNG Studies, Boroko, 1980, pp. 247-248.

[8] Ibid, p. 4, and interview with Nitsche and Murray, 20 September 1979.

[9] Some academic observers are particularly inclined to minimise the extent to which individuals and groups in Papua New Guinea are gaining control and ownership over land and utilising it for commercial purposes. Thus, Chris Gregory has asserted, from a purportedly marxist but static perspective, that 'land has not yet emerged as a commodity on any significant scale', but has made no effort to examine the actual process of emergence, the developing commercial usage of land in the most advanced regions of capitalist agricultural production. With even greater detachment from reality, he also claimed that 'there is a tendency for commercial land to revert to clan land as well as a tendency for the equalisation of land distribution between clans'. Chris Gregory, 'Papua New Guinea', Bowyang, 4, 1980, p. 109.

[10] Alan Ward, 'Customary Land, Land Registration and Social Equality', in Denoon and Snowden, op. cit., pp. 252-258, 260-264.

[11] Land Disputes Settlement Act 1975: What It Is and How It Works, Prepared and printed by the Office of Information, Konedobu, 1977, p. 1.

[12] While freehold ownership was by no means entirely absent either, as chapter 4 above has indicated, the reality of land usage did not dichotomise (as did Gregory and others) between communal land ownership and land as a commodity, between the absence of freehold and freehold.

[13] Department of Lands, Surveys and Environment, Mobilising Customary Land for Larger Business Proposals, Boroko, n.d.

[14] Ibid, p. 2.

[15] Ibid, pp. 3-4.

[16] Ibid, p. 5.

[17] Ward accurately observed that 'land registration is an expensive and rather complex process, favouring the elite', op. cit., p. 261. See Good and M. Donaldson, Development of Rural Capitalism in Papua New Guinea: Coffee Production in the Eastern Highlands, Institute of PNG Studies, Boroko, 1980.

[18] Personal interview with Murray, 27 and 28 November 1979.

[19] Land Disputes Settlement Act 1975, p. 9.

[20] Personal interview with Murray, 27 and 28 November 1979.

[21] Personal interviews 31 August 1979; 20 September 1979; 27 and 28 November 1979; and Post-Courier, 23 August 1979, and Dennis Appendix, op. cit.

[22] R.G. Ward, et. al., Growth Centres and Area Improvement in the Eastern Highlands District: A Report to the Central Planning Office of Papua New Guinea, Canberra, ANU, Department of Human Geography, Research School of Pacific Studies, 1974, pp. 18 and 23; good and Donaldson, op. cit., p. 30.

[23] Personal interviews 20 September 1979; 27 and 28 November 1979.

[24] Post-Courier, 31 August 1979, p. 1 and figures supplied by the Coffee Industry Board, Goroka, 27 November 1979; and Dennis Appendix, op. cit.

[25] Material in this section is based upon personal interviews as for [23] unless otherwise noted.

[26] Personal interview 27 November 1979 with Murray and applicant.

[27] The small class of rural capitalists were significantly differentiated from the larger rich peasantry in their greater control and ownership of resources and usage of wage labour. See chapter 4 above.

[28] Good and Donaldson, op. cit.
[29] Information supplied by Kevin Murray, DPI, Goroka, 27 November 1979.
[30] Personal interview with Murray, 27 and 28 November 1979.
[31] D. Anderson, An Economic Survey of Smallholder Coffee Producers - 1976, DPI, Port Moresby, September 1977, p. 16.
[32] Personal interview with Murray, 27 November 1979.
[33] Personal interview with Murray, 27 November 1979.
[34] Op. cit., pp. 16, 22.
[35] 'Information Paper ...', p. 4, personal interview 27 and 28 November 1979, and Dennis, op. cit.
[36] Personal interviews with Nitsche and Murray, 20 September 1979; 27 and 28 November 1979.
[37] Post-Courier, 22 January 1980, p. 14.
[38] Very roughly, one hectare could be expected to produce 2.2 tonnes of coffee and, at an estimated price of K1,350 per tonne, 20 hectares would return K59,400 gross. Running expenses would be K32,000 after full production and some K27,000 nett might therefore be anticipated.
[39] Village Economic Development Funds allocated to the Eastern Highlands, September-December 1978, totalled K88,400, and in the first six months of 1979, totalled K77,800. Post-Courier, 5 September 1979, p. 24.
[40] Personal interviews with Nitsche and Murray, as for [36].
[41] A.J. Kimber, 'Current Recommendations in Arabica Coffee Management', mimeo, Coffee Industry Board, Goroka, n.d., p. 1.
[42] A.J. Kimber, 'Future Research Work on Arabica Coffee', mimeo, Coffee Industry Board, Goroka, n.d., p. 1.
[43] The following section is based on a personal interview with Godfrey Benjamin, Linupa's Business Manager, Lantern Lodge, Goroka, 18 September 1979.
[44] Benjamin referred to Barry Holloway and Gai Duwabane, personal interview 18 September 1979.
[45] The following section is based on a personal interview with Craig McConaghy, Executive Officer, Provincial Secretariat, Goroka, 19 September 1979. Further detail is found in Good and Donaldson, op. cit.
[46] Turnover for Collins and Leahy Holdings Ltd ranged from K8 to K10 million in the early 1980's. They continued a heavy capital expenditure programme and opened a new department store in Goroka in June 1981, Australian Financial Review, 1 October 1981, p. 22.
[47] The following section is based on a personal interview with Anders Berqvist, Gouna Manager, Gouna Offices, Goroka, 19 September 1979. Further detail is found in Good and Donaldson, op. cit.
[48] About a year later its helicopter fleet totalled ten machines and indigenous ownership reached 70 per cent. Post-Courier 29 January 1981.
[49] The following is based on a personal interview with Akepa Miakwe and Paul Asaro (Secretary of Bena), ANGCO Offices, Goroka, 19 September 1979. Further detail is found in Good and Donaldson, op. cit.
[50] Personal interview 20 September 1979; and 'Five Year Programme for Commercial Vegetable Production 1980-84'; and 'Five Year Programme for Subsistence/Nutrition 1980-1984', mimeos, DPI, Goroka, n.d.

[51] Personal interview with Murray, 27 and 28 November 1979.

[52] R.F. Salisbury, 'Changes in Land Use and Land Tenure Among the Siane of the New Guinea Highlands', Pacific Viewpoint, 5, 1964, p. 8; and G. Harris, 'Cash Cropping or Subsistence Farming - A PNG Dilemma', Pacific Islands Monthly, 49, August 1978, p. 63.

[53] Personal interview with Murray and Nitsche, 20 September 1979.

[54] Bureau of Statistics, Rural Industries 1975-76; Port Moresby, 1978, Table 27.

[55] Personal interview with Murray and Nitsche, 10 September 1979.

[56] For a broad consideration of plantations in PNG, some aspects of which are unfortunately utopian and rather passe, see Michael Walter, What Do We Do About Plantations?, Discussion Paper No. 34, Institute of Applied Social and Economic Research, Boroko, November 1980.

Conclusion The successful articulation — its scope and limitations

The articulation of non-capitalist and capitalist agricultural modes is often a slow and difficult process, but in the highlands of PNG coffee production for export was successfully implanted in some twenty years. When coffee production was seriously initiated in 1950 the country exported 33 tonnes; in the 1960-1961 coffee year (October to September) 2,080 tonnes were exported; in 1976 40,000 tonnes; and in 1980 exports totalled 51,037 tonnes. The expansion of output was around 11 per cent annually through the first half of the 1970s.

Table 7.1
Coffee exports

	Quantity (tonnes)	Value (Kina m)
1950	33	--
1960-1961	2,080	--
1976	48,000	100.16
1977	36,965	143.44
1978	46,125	107.25
1979	49,585	125.00
1980	51,037	118.69

Source: Donaldson and Good, 1980, pp. 164-165; P.N.G. Bureau of Statistics, Abstract of Statistics, March 1980 and March 1981.

Coffee became PNG's most valuable export in the latter 1960s, and held this position until the Bougainville copper mine under its parent company Conzinc Riotinto came into production in 1972. But coffee remained easily the most valuable agricultural commodity holding the dominant position within the 'Three C's' which form the bulk of agricultural exports.

Table 7.2
Value of exports by type of commodity (Kina m)

	1976	1977	1978	1980
Coffee	100.16	143.44	107.25	118.69
Cocoa	39.13	86.34	62.96	46.49
Copra	12.13	22.96	23.02	24.59

Source: Abstract of Statistics, PNG Bureau of Statistics, June 1978, December 1979 and March 1981.

Even after the turn to copper, coffee's importance as a revenue earner was outstanding. In 1976 coffee exports represented 25.6 per cent of the country's total exports, and in 1977 they were worth (at K143m) 70 per cent of the value of copper exports (K201m). [1]

Coffee growing quickly became an activity wherein indigenous producers predominated. The participation of indigenous farmers in coffee outstripped that of any other commercial sector. This was the case with regard to both the area under the crop and to the share of production. In 1972-1973 the area held by indigenous farmers represented 76 per cent of the total for coffee (compared with 56 per cent for coconuts and only 29 per cent for cocoa), and the proportion of indigenous growers in coffee production was 72 per cent (compared with 40 per cent for coconuts and 31 per cent for cocoa). [2]

The involvement of Papua New Guinean farmers in coffee has significantly increased during the 1970s, through (as noted in chapter 6) the substantial acquisition especially in the Eastern Highlands of foreign owned plantations and the beginnings of the consolidation programme. Originating in that province, commercial coffee production spread westwards, with Goroka remaining at the centre of what was basically a highlands industry. [3] Of PNG's total production of green bean in the coffee year 1975-1976 of 43,003 tonnes, 37,215 tonnes were produced in the highlands, principally in the Eastern Highlands, Simbu and Western Highlands. [4] Near the end of the 1970s, some eighty per cent of the adults in the Eastern Highlands were engaged in coffee growing, and there was little good land which was not under the crop. [5] Coffee has achieved a position of great importance in the national economy and among the great bulk of highlands peasants in quite a short time.

The people who had taken up the new cash crop were obviously not the mere 'stone-age cultures' who lived in a land 'where time stood still' as foreign misconceptions too frequently presented them. The extent, strength, and speed of the transition to coffee was based upon the receptivity of the non-capitalist systems, which had much earlier undergone the transition from yam and taro production to the higher productivities of sweet potato. The dynamic and hierarchical elements in advanced sweet potato- and pig-producing systems were clearly personified in the powerful and ambitious big men. Their position was influenced by war and trade, but was most directly related to the exploitation of the agricultural labour of women, in some places also of 'serfs'. These non-capitalist agricultural systems drew the admiration of the first whites who entered the highlands. In the early transactions between the agents and fore-runners of the new colonialism and capitalism, on the one hand, and the non-capitalist societies, on the other, it was the big men who were prominent. They not only exercised political leadership, significant in conditions of war and conquest, but also controlled access to the region's agricultural surplus, and to labour and land, the keys to development in the long term.

Australian colonialism did not destroy the powers of big men. An important consequence of the colonial peace and the penetration of capitalist trade was the increased availability of men's labour. Good land was also obtainable in the central cordillera, and the highlands labour power and fertile valley grasslands, which so attracted colonial capitalists at the end of World War Two, were factors firmly under the jurisdiction of big men. Ambitious big men were the leading and outreaching elements in advanced agricultural systems with already well developed mechanisms of exploitation. Coffee was an appropriate rich cash crop in the broadest sense, climatically, topographically, and socially. The cultivation of coffee, most significantly, could promise to transform the position of big men, to provide much greater wealth and power, if they were among those of their kind who could see that coffee really was, as the phrase had it, 'the money which grows on trees'.

Agricultural capitalism, through the agency of Australian colonialism, meshed well with the non-capitalist systems of the highlands after 1950. Australia did not itself produce coffee, and the colonial state in PNG could obtain desirable revenue as well as a possible bulwark in the form of rising rich peasant coffee growers for the plantation system, then sadly representative of commercial agriculture in PNG and typified by unattractive and moribund copra on the coast and islands. [6] The spectre of peasant rebellion in the 'White Highlands' of Kenya, as we have seen, over-exercised the minds of foreign coffee planters in the highlands of PNG in the 1950s, and the rise of a man like Sinake Giregire, an early rich coffee farmer and vocal defender of large scale commercial agriculture, was partly a consequence of this. After the very brief but intense period when many foreign owned coffee plantations were established especially in the Goroka area, state assistance for commercial agriculture was largely directed towards the peasantry. Given the organisation of the non-capitalist societies and the institutions and values of capitalism, big men who typically would hold other state offices or employment by this time, were the intermediaries between the two systems, and eventually the greatest beneficiaries of the peasant based developmental programme. The role of the foreign owned plantations had some similar effects, in the patron-client relations developed between some thoughtful planters and active village headmen, with the latter assisting in the provision of the essential land and labour, and then profiting most directly and individually from the well known 'demonstration effects' of the operational plantation. Certain _kiaps_-turned-planters, with a close knowledge of highlands societies, established mutually advantageous relationships where a valued senior employee (typically _bosboi_ or foreman) was materially assisted in establishing his own coffee farm. Foreign coffee plantations were not merely or totally exploitative, and certain of them gave something back to a few of the people among whom they were located.

But it was the agricultural extension services of the state which contributed most in the long term to the diffusion of coffee among the peasantry and so to the rise of a rich peasantry out of the ranks of the big men (and overlapping colonial triad of _luluai_, _bosboi_, and cathecist). [7] When the colonial state also involved itself in the

1960s and 1970s in other developmental activities, through the formation and operation of rural cooperatives, the promotion of cattle raising, the establishment of titles to land and, most significantly, the provision of bank credit for commercial agriculture and ancillary businesses, commercial agriculture became very attractive indeed to much of the highlands peasantry. The wider diffusion and strengthening of commercial agriculture among highlanders was accompanied simultaneously, however, by the greater concentration of key developmental resources among a small and leading minority. They alone tended to obtain organisational experience, enlarged their coffee holdings perhaps combining these with cattle, gained title to land, and began to acquire bank credit for what were in some cases increasingly diversified capitalist operations. These results were brought about by the following elements separately and in combination: state support for the development of peasant based commercial agriculture; an appropriate and attractive crop in the origin and at the centre of the whole trajectory; and non-capitalist societies which were sufficiently dynamic and hierarchical to transform themselves into a differentiated peasantry.

An important absence in the approach of the colonial state to coffee production by highlands peasants was coercion. Other colonial systems, in different places and times, had of course used force upon the colonised with, in some cases, very negative, long term results for agricultural production. The Portugese in Mozambique enforced production from the mid 1920's of cotton, a labour intensive crop competing directly with food production. Zones were established wherein each household had to cultivate 1 or 0.5 hectares of cotton and the produce was compulsory taken over at a fixed price by companies with territorial concessions. The peasantry was also subjected to forced labour for the production of such cash crops as sisal, sugar, tea, coconuts and cotton, on the plantations and large estates of foreign commercial interests and Portugese immigrants. Colonial coercion of the peasantry actually affected 'the large majority of the people', who were 'forced to work for extremely low wages'. In consequence, by the 1970s 'the notion of work had become almost synonymous with the notion of exploitation', a serious outcome for both individual and social production in the independent country. [8]

Australian colonialism used coercion over copra production in coastal areas earlier in PNG, and compulsion was widely used to obtain village labour for such things as road and bridge construction and maintenance. It was one of the frequent tasks of the luluai (before and after independence) to assist the kiap in mobilising this labour. But state coercion over coffee production only existed in the attenuated form of the lack of any similarly attractive alternative to coffee. Highlanders accepted the production of coffee with enthusiasm, on the basis of their own socially differentiated initiatives and responses.

Another aspect of state assistance for coffee, in particular, and commerical agriculture generally, was itslargely direct relationship to peasant production. Vital services and resources, various extension activities, land titles, bank credit, were made available by the state directly to the successful peasant producer in coffee, cattle and closely associated commercial activities like transportation. There was little wastage of state funds and scarce facilities on the 'special projects' which have sometimes characterised and ultimately limited the scope and effectiveness of agricultural development in other places. The non-existence of foreign aid agencies in agriculture in PNG, between 1950 and 1980, also contributed to the coherence of development policy, and the transition to commercial agriculture was facilitated by the specificity and directness of the orientation of the state. [9]

The success of the articulation in PNG is the more outstanding when contrasted with the failures to establish commercial agriculture among peasantries in other places, such as peasant based coffee production in the inter-War years in Tanzania [10] where many favourable factors existed. The Chagga of the Kilimanjaro district possessed good volcanic soils and a highland climate suitable for coffee, they already had a 'sophisticated agriculture' with available surplus labour, railway transport, and the possible model of foreign owned plantations located in the district. [11] Early Chagga growers had produced coffee in German times, in 1916 they owned some 100,000 trees, by 1925 some 6,700 farmers possessed about 990,000 trees, and in 1930 the total reached nearly 6 million coffee trees. In 1925 the Kilimanjaro Native Planters Association was formed, and Chagga farmers were buying and selling land in 1927. But the foreign planters had earlier begun to demand that the Chagga should be forbidden to grow coffee, hence the better obliged to work as wage labourers for them, and in 1927 the colonial state decided to discourage African cultivation of the high value arabica coffee variety. [12]

Buhaya peasants near Lake Victoria took to coffee with even greater enthusiasm. [13] They planted 2 million coffee trees during the year 1923, and by 1928 more than half the taxpayers in Bukoba district grew coffee and exported, in that year, 7,973 tonnes (cf. 314 tonnes sold by the Chagga). No foreign planters existed in this area. The Buhaya social formation possessed hierarchical elements, wage labour was more common here than in the Kilimanjaro, and land sales advanced further to be 'common' by 1924. [14]

Despite the fact that the movement to cash crop production among both the Buhaya and Kilimanjaro was supported by a 'powerful thrust of indigenous entrepreneurship', [15] no permanent articulation took place. Weaknesses present within the pre-capitalist societies would seem to have played a part in this negative outcome. Old hierarchies, where they existed, were not necessarily interconnected with non-capitalist production dynamics and newer entrepreneurial tendencies. Chiefs benefitted from contacts with whites and were favoured by the colonial state in the introduction of cash crops; but chiefs, said Iliffe, 'found difficulty in controlling commercial agriculture' and were 'seldom practising cultivators'. Unlike the PNG highlands big men whose whole raison d'etre was deeply implicated in agricultural production, chiefs in the favoured cash crop areas of Tanzania sometimes actually disdained cultivation. [16]

Such an internal impediment was very serious, but its effects on the articulation were compounded by increasing opposition to peasant cash crop production from within the new colonial capitalist system. First came the active hostility of foreign planters in some areas combined with a readiness to resort to colonial compulsion, and then a reduction of state support for peasant coffee. In the 1930s possibilities for peasant cash crop propsperity in Tanzania faded. Britain's economic position was weakening, and Tanzania was largely abandoned by British investment which had far more attractive colonies to consider. Despite early positive factors and achievements, [17] articulation had foundered in Tanzania after two to three decades, with serious and long lasting consequences.

Other examples of the problematic position of peasant agriculture include the many cases where articulation never seriously developed in the first place, even over very lengthy periods. In situations such as those of Zambia and Mozambique, largely foreign owned systems of capitalist agriculture were simply installed in the countryside and built upon the land, labour and other resources of the indigenes. In colonial Mozambique such exploitation was both persistent and heavy but it was also very evident in Zambia. The subsequent limited development of capitalist agriculture was contingent upon the persistent under-development of peasant farming and rural communities in these countries. [18]

The failed and the uninitiated articulations are, however, basically similar for in both cases no cooperative and dynamic inter-relationship came into existence between the capitalist and non-capitalist systems, and the development of the former was predicated upon the containment or destruction of the latter. Commercial agriculture did not work with and grow out of the productive capacities and tendencies of the non-capitalist systems. It is not the presence of an 'uncaptured peasantry', homogenous, autonomous, inherently opposed to capitalism, [19] which defines the backwardness of agriculture in many places today, but rather the creation of, as it were, 'unarticulated', repressed, contained, ignored and discarded peasantries. Backward agriculture occurs when earlier agricultural capacities were halted or stalled, where existing potentialities for comercial development were unassisted in appropriate and meaningful ways, where perhaps, as in Zambia and Mozambique, a peasantry's resources of land and labour were all too extensively 'captured' by colonial capitalism earlier. The situation of PNG contrasts sharply with this. In the highlands of PNG highly appropriate forms of assistance were extended by the colonial state to the hierarchical and dynamic forces of non-capitalist agriculture, and out of a cooperative engagement between the different systems successful coffee producing peasants enthusiastically arose.

Class action by rich peasants and subsequently also by the smaller class of rural capitalists in PNG made a powerful contribution to this development. Their activities were broad and consequential. They sought elected office and influence in the leading colonial and post-colonial political institutions, councils, national assembly, provincial government, as and when these bodies appeared, skilfully manipulating the fact that they were simultanteously the peasantry's established and 'natural' leaders and the exemplars of new commercial, entrepreneurial abilities. In the Eastern Highlands the rich classes gained a preponderant influence in the main councils and provincial government. The modern political institutions were utilised to advance their individual and class interests, as similarly ambitious big men, in looser and more individuated ways, had profited from the offices of luluai, bosboi, and catechist. In relation to the poorer peasants, the rural rich were particularly concerned with capturing governmental resources through the provincial government and councils, but they were also concerned about such issues as the use of the state's legal and police powers against banditry and other threats to property. In their relations with the foreign bourgeoisie in agriculture and commerce in the highlands, the indigenous businessmen were usually more cooperative than competitive, concentrating on the steady acquisition of coffee plantations and trading companies as the foreigners willingly departed through the 1970s.

The outcome of this political and economic action vis-a-vis poor peasants, the foreign bourgeoisie and the state, represented a large expansion and gradual diversification of the property and economic activities of the rich, and crucial further impetus to highlands capitalist development. This was most notable in the establishment and growth of the devlopment corporations, first by the leading rural capitalists of the Eastern Highlands and then by the provincial government and richest councils. The corporations mobilised villagers' savings and support, organised managerial skills, and attracted investment and loan funds for the expansion of indigenous economic ownership. Not only was the scale of coffee production by highlanders increased, but control over marketing was also obtained, along with a widening involvement in commercial agriculture and commerce generally. A remarkable cohesion among the small group of leading rural capitalists strategically located in plantation production, commerce, the development corporations and key political institutions in the Eastern Highlands in the latter 1970s, contributed to the strength, direction and coherence of the development process. The rich rural classes rising out of the non-capitalist formations were the strongest indigenous motor of capitalist agricultural development in PNG.

The limitations of the articulation

The social costs of the successful articulation are nowhere more evident than in the increasing inequalities between a small minority of rich peasant and rural capitalist farmers, on the one hand, and the vast majority of very small coffee farmers, on the other. The inequalities of the old societies have been extended and deepened within the new capitalist system. As discussed above, most coffee growers control little more than a very small amount of land and a few hundred trees. Their position actually worsened through the 1970s with the average Eastern Highlander holding just 0.1 hectares with 250 trees in 1976, or in another estimate of the same year, over half the growers surveyed in the Eastern and Western Highlands and Simbu held less than 500 trees. In sharp contrast, the same survey found that only 6 per cent of farmers possesed more than 1,000 trees.

While the majority held little land and few coffee trees and almost nothing more, the small minority owned large coffee farms, the ten Gorokan business leaders of the late 1960s owned an average of 9,000 trees each, and much else besides. They owned or had access to, all or most of a wide range of productive resources: wage labour (an increasingly important factor in the minority's success); ownership of equipment; access to loan funds (impossible for the majority); access to extension services (increasingly concentrated upon the minority); and the possession of management skills (an important attribute of the existing plantations and of the new consolidated coffee farms). These factors represented wealth and income in themselves and they also opened opportunities, or strengthened opportunities already taken, for political office, organisational position and the diversification of commercial activities. The wealth and power that was open to the small minority was suggested by the position of Akepa Miakwe in the mid-1970s. He had some 10 hectares of coffee with about 7,000 trees, two coffee pulpers, a commercial vegetable garden, trucks and a trade store, and among other things he was, as considered above, the director and virtual founder of the Bena Development Corporation. Also a member of the national assembly since 1972, he claimed an annual income in 1976 of some K30,000. [20]

Such attributes were not dissimilar to those of Hari Gotoha, [21] and Auwo Ketauwo, and certain other individuals referred to above, and were in fact typical of the wealth and power of the tiny group of highlands rural capitalists. The consolidation programme beginning at the end of the 1970s would further extend their personal wealth, not least in regard to the new coffee varieties which would permit far greater densities of coffee trees, up to 10,000 trees per hectare on perhaps a 15-20 hectare farm, and the greater emphasis on the provision of management skills and of loan funds by the state.

In contrast with the increasing wealth of the rich rural classes, the position of the average coffee grower in the Eastern Highlands deteriorated. An average holding of 0.1 hectares in 1976 produced about 74 kilos of coffee for the farmer, although in 1973 an average holding of 0.15 hectares and some 360 trees had produced 152 kilos per grower. [22] Over the period 1972 to 1976 world coffee prices had fluctuated between K751 and K2,663 a tonne. An official calculation suggests that in 1976 the average 'smallholder' income per head from all export crops in the Eastern Highlands was about K30 only. [23] The average coffee grower's income (coffee being a rich cash crop) was larger than this. At the highest of the above world coffee prices, K2,663 a tonne, the average farmer's output of 74.2 kilos in 1976 represented a gross income of K199.82 for the grower and at the average price, K1,707 a tonne, it offered an income of K126.66. [24] The average coffee grower's net income after expenditures would have been much less. In either case, the gap between the wealth and power of the rich rural classes and the position and income of the ordinary peasant coffee grower was extremely large.

Similar inequalities characterise the development of cash crop production in other places. Among the Chagga in Tanzania in 1930 only one man in three grew coffee, but of those who did, 96 per cent owned less than 1,000 trees, normally grown inter-cropped with their bananas, while 4 per cent owned a hectare or more of coffee, which they often grew in distinct plantations and with hired labour. (One of the largest, in 1932, owned over 12,000 trees.) The Matengo people in turn adopted coffee from the Chagga and the determination of their entrepreneurs was such that in 1937 four-fifths of all Matengo coffee was produced by only four out of a total of 743 growers. [25] The long devotion of Ghanaian peasants to cocoa resulted in particularly severe inequalities, the distribution of money income among cocoa farmers became such that, in 1963-1964, while the share of the lowest 20 per cent of farmers was only 2.4 per cent of the total accrued income, that of the highest 20 per cent was 56.4 per cent. However, by 1970, these already severe inequalities had worsened, the lowest 20 per cent of farmers receiving only 1.8 per cent of total cocoa income, while the highest 20 per cent of growers obtained 62.5 per cent. [26]

The situation of Ghana was rivalled by that of the Ivory Coast where both rich cash crops, cocoa and coffee, are grown. The variations in total cash income per farm in 1978, were very great, the earnings of farms over 20 hectares in size being almost 35 times larger than those under 2 hectares. [27]

If the implantation of rich cash crop production occurs best where old agricultural surpluses and hierarchies existed, the successful development of these commercial systems is certainly accompanied by increasing economic inequalities amongst the peasants. More accurately, the growth of rich cash crop production partly originates through the economic drive of embryonic rich peasants who accumulate resources as commercial development proceeds. The inequalities in the production of rich cash crops in Ghana and the Ivory Coast in the 1960s and 1970s appear broadly comparable to those in PNG at the time. The development of capitalist productive forces is accompanied and accelerated by the development of capitalist social relations, in agriculture and with peasants, as in manufacturing and mining with the working class.

Women agriculturalists have been seriously disadvantaged by the development of coffee production in PNG, but a very small minority has been affected in opposite ways. Their position as food producers in the non-capitalist societies has been maintained in the transition to commercial agriculture while the activity itself has been subordinated to cash crop production. The sphere of food production for home consumption has been more strictly and completely relegated to women, while cash crops are largely the preserve of highlands men. Adult women in the 1970s spent far more time than men in family food production, in which highlands men have become increasingly disinterested. [28] The state has provided little assistance to the commercial production of food and far less to food for subsistence usage, despite their impact on the national economy which valued staples for family production at approximately K240m in 1976, and marketed domestic food production at an additional K30m. [29] Domestic food production continues without benefit of financial or technical assistance, on the basis of the increased labour inputs of women and help from children.

In addition women and children provide the bulk of the labour for the most arduous tasks such as weeding and picking. From the 1960s there were 'significant' numbers of women working as casual labourers on coffee plantations, and their position in general was that of a very low paid, seasonal or reserve labour force. [30] While becoming casual labourers in coffee, women of necessity have also attempted to become the marketers as well as the producers of food, as extensions of their entrenched subsistence activities. Despite the fact that sales of food at the two main highlands markets in Goroka and Mt. Hagen in 1976 were estimated to value K2m each, the daily returns to the vendors, in large majority, women, were low. [31] Overall, the value of women's productive activities within commercial agriculture had clearly increased, without any commensurate return or social recognition. 'The steadiness of domestic production by women', as the Enga Province Rural Development Study Group expressed it in August 1978, [32] buttressed the highlands' economies. The varied forms of this productive labour represented an extremely heavy burden on women.

Some women have lately begun to grow coffee themselves, making up about 3 per cent of all highlands growers in 1976, but their plots tend to be even smaller than average. [33] The only real exceptions to the suppressed position of women in agriculture are the wives of the leading members of the rich rural classes, their few counterparts among educated, professional highlands women in urban centres, and a fewexceptional women entrepreneurs. Their presence was indicated by the formation of the Goroka Women's Investment Corporation in 1975, with help and encouragement from the state. Shares in the corporation were K50 each, placing them well beyond the reach of the average marketer let alone ordinary village women, and most shareholdings initially came from women in the developed coffee growing areas around Goroka assisted in some cases by their already successful, perceptive husbands. These were the districts from which the established Gouna Development Corporation drew its membership, and the Goroka Women's Investment Corporation was run on similar lines to Gouna in the 1970s. Its first venture was the opening of a coffee shop at Goroka airport, and it subsequently entered commercial agriculture with a peanut plantation on land leased to it by the state. It was an employer of wage labour, and its functions were not dissimilar to those of the Bena and Gouna corporations in matters other than scale. The chairwoman of the Goroka Women's Investment Corporation's board of directors was Betty Ketauwo, the wife of the chairman of the Gouna corporation, and leading rural capitalist, Auwo

Ketauwo. [34] The activities of this elite group of women were unlikely to alter the structural subordination of women under capitalist agriculture but could well offer additional reinforcement to the rural rich.

The successful development of coffee production had brought further economic costs in the late 1970s when, as noted above, almost all the best land in the Eastern Highlands had been given over to coffee. In the view of Harris, coffee was 'a relatively heavy user of land', [35] and the new consolidation programme promised to place even unused hill-side and gully land under the crop. The production of coffee under the sway of the leading indigenous growers would be further advanced, but with increased difficulties for food production and the women agriculturalists already struggling in this sector. [36] The successful development of commercial agriculture in PNG had also been very uneven geographically. Nationally, it was largely confined to the production of cocoa in the islands and coffee in the highlands, and at the regional level the concentration of coffee was uneven too. The development of coffee and the growth of the infrastructure, towns and administrative centres of the highlands had gone on together, with the distribution of the crop influenced of course not only by topography but also by whether or not a particular community possessed a big man who saw the wealth which coffee represented. Coffee was centred around the main towns of the Eastern and Western Highlands and Simbu and areas adjacent to the Highlands Highway. Coffee, even more so than sweet potato earlier, had divided the highlands into areas of advanced and of retarded agricultural development. The remote and higher areas of the Eastern Highlands had become an area of backward rural capitalism, and large parts of Simbu had served to provide labour for plantations elsewhere. The development of cocoa had been rather similarly concentrated in certain parts of Ghana, gradually disadvantaging other regions and peoples in the process.

A realistic assessment of the success and limitations of the establishment of coffee production in PNG must be made, however, in the light of the actual and possible alternatives. As regards productive activities of comparable scale and value, rich cash crop production firmly gripped by indigenous growers seems preferable to the strategy of minerals' exploitation by foreign multinationals. The production of copper and gold as it is carried on at Panguna on Bouganville and at Ok Tedi is large scale, capital, and technology intensive, and it is of limited, even uncertain, duration beyond the short term. More particularly, it offers a very limited role in ownership and control for indigenous classes. All that it really offers in this respect is minority participation by the state, that is, a limited role by an indigenous administrative or petty bourgeoisie not itself involved in production.

While copper and coffee are commodities similarly affected by sometimes widely fluctuating world prices, coffee has the basic distinction of being a resource which is relatively accessible to producing classes in PNG. From their successful involvement in coffee, the rural rich have therefore been able to carry out such innovations as the formation of the development corporations, modern organisations which have grown out of coffee but where the leadership of the rural rich transcends their old personal linkages with clans and villages and which expand and diversify indigenous economic activities in commercial agriculture, transport, marketing, and to a limited extent, manufacturing. Additionally, coffee is also important to highlands poor peasants as a comparatively independent income generating activity. Unlike other agricultural commodities such as copra or tea, coffee can be grown on small areas of land with low labour inputs. The position of the average highlands grower with, in the late 1970s, only about one-tenth of a hectare and an annual output of some 75 kilos, is not to be properly understood simply in comparison with the large holdings and output of the rich coffee farmers. In comparison with low paid plantation labour; with a very low value and labour intensive crop like copra; with the inbuilt dependence of the smaller tea grower on a nuclear estate; with the struggles particularly of women, to produce food commercially; and with the hazards of an existence based on subsistence, small scale coffee growing was attractive.

It is relevant that in the latter 1970s there was then no notable exodus from the land to squatter settlements in the towns in the areas of advanced cash crop production, the outflow of wage labour from Simbu being part of a long existing trajectory which assisted the advanced regions. The coffee consolidation programme could well produce squatter settlements in the Eastern Highlands in the future, but as yet these were a feature precisely of the stunted agriculture of the coast and associated enclaves like Port Moresby and Lae. Rich cash crop production by peasants was the best of the actually available development strategies in PNG, despite the limitations which it brought in its train. Egalitarian cooperative agriculture was not present in the past non-capitalist systems, and no viable non-commercial agricultures were offered by the colonial and post-colonial state.

If the economic rise of rich peasants in PNG was the unavoidable concomittant of peasant-based commercial agricultural development under favourable circumstances, the extension of their political power today over the poorer peasants is a somewhat different matter. The fairly widespread attacks on property and land and other disputes in the highlands in the latter 1970s were indications of deep discontent among the mass of the population. Poorer peasants, however, faced sharpened dangers in the 1980s as the rich rural classes extended and consolidated their control over increasingly scarce productive resources. The situation was worsened and indeed defined by the continuing absence of any independent political organisations of the poor, which ensured that deprivations and even widespread resentments remained dissipated in banditry and other inchoate actions. But the formation of modern political organisations of the poor peasants was unlikely to be achieved by themselves alone. Provincial and national government offered economic and political support to the highlands rich rural classes in established fashion. Any organisational and political support to the poor peasants could only derive from classes or class segments which were independent of the state and the rural rich. The political emancipation of the poor peasants could be approached in organised collaboration with either sections of a middle peasantry with a stronger foothold in coffee or an educated petty bourgeoisie or working class from the towns. This itself would be a long and difficult road. The creation and political subordination of the poor peasantry was one of the gravest limitations produced by the articulation of non-capitalist and capitalist agriculture in PNG.

Notes

[1] Donaldson and Good, 'The Eastern Highlands: Coffee and Class', in Donald Denoon and Catherine Snowden (eds.), A Time to Plant and a Time to Uproot, Boroko, Institute of Papua New Guinea

Studies, 1980, pp. 164-165; Abstract of Statistics, PNG Bureau of Statistics, March 1980.

[2] World Bank, Papua New Guinea: Economic Situation and Development Prospects, Report No. 1150-PNG, July 14, 1976, Annex 1.

[3] Goroka is the headquarters of the Coffee Industry Board among other things.

[4] D. Anderson, An Economic Survey of Smallholder Coffee Producers - 1976, Port Moresby, Department of Primary Industry, September 1977, Table (i).

[5] See chapter 6 above.

[6] Cocoa in the islands of PNG, another appropriate rich cash crop in a receptive non-capitalist society, played a similar but less significant role as the statistics quoted immediately above suggest.

[7] As shown in chapters 3 and 4, a well-placed big man was likely to be also a luluai or hold certain other colonial office or employment in the late 1940s and 1950s. Soon after, other modern political offices, particularly local government councillor (and the position of council president especially), became of key importance to the rising rich peasant.

[8] R.K. Srivastava and I. Livingstone, 'Growth and distrubution: the case of Mozambique', in Dharam Ghai and Samir Radwan (eds), Agrarian Policies and Rural Poverty in Africa, Geneva, International Labour Office, 1983, pp. 253-255 and 269-270.

[9] Zambia exemplifies the situation of the proliferation of both 'special projects' (e.g. land settlement schemes, state farms, producer cooperatives, so-called rural development centres and national service schemes) and foreign agencies in agriculture and, in direct consequence, of the dissipation of available investment resources and of increased incoherence in state support for agriculture. William Smith and Adrian Wood, 'Patterns of Agricultural Development and Foreign Aid to Zambia', Development and Change, 15, 3, July 1984; Good, 'The Reproduction of Weakness in the State and Agriculture: Zambian Experience', African Affairs.

[10] John Iliffe, A Modern History of Tanganyika, Cambridge, Cambridge University Press, 1979, chapter 9, which forms the basis of the following section.

[11] Ibid., p. 274.

[12] Ibid., pp. 275-277.

[13] Hyden says that coffee was 'traditionally grown' on an intercropping basis with bananas by the royal clans in Buhaya, and German colonialism introduced commercial coffee growing among the peasantry in 1904. He gives an impression that this was and remained, the poorer quality and lower valued robusta variety. Goran Hyden, Beyond Ujamaa in Tanzania, Berkeley and Los Angeles, University of California Press, 1980, pp. 49-50.

[14] Iliffe, op. cit., pp. 281-283.

[15] Ibid., p. 290.

[16] Ibid., pp. 290-291.

[17] It is noteworthy that Hyden's account of peasants in colonial Tanzania differs from the above in terms of his first principles, which tend to assume 'the conflict between the [capitalist mode] and the peasant mode', and which declare that 'peasants wanted access to modern goods and services without sacrificing the economy of affection'. This is of course consistent with Hyden's espousal of the 'inward approach' perspective on peasant studies,

outlined above in chapter 1. In order to maintain his position however, Hyden has to ignore and heavily qualify much relevant evidence which he himself presents. As regards the Chagga he has this to say:

> The creation of estates accelerated the pressures on the land in Kilimanjaro and forced the Chagga peasants to be more outward looking. They were generally more open to advice concerning modern farming ... To some extent, entrepreneuralism was the product of the indigenous social system among the Chagga people. (Our emphases)

He concluded his account by suggesting that 'nowhere' in colonial Tanzania was 'petty-capitalism' more 'easily developed' than among the Chagga, although Iliffe's analysis indicated that the Buhaya peasants accepted coffee production with equal if not greater enthusiasm and consequences through the 1920s. Hyden is able to maintain his position and his commitment to the idea of an homogenous peasantry devoted to its 'economy of affection' largely through his careful choice of terminology, his reference to the appearance of 'petty-capitalism' in the countryside rather than the development of rural capitalism or commercial agriculture, and his related emphasis on the 'emerging African petty-bourgeoisie', detached from and quite different to the mass of the peasantry (from whose ranks they have nevertheless come), and his avoidance of the concept of formation of a rich peasantry. Hyden concluded his account of commercial agricultural development in colonial Tanzania with the statement that 'the British had set in motion a process over which they lost control', which of course totally ignores the strong, varied and sometimes originating contributions from peasantries in the areas of relatively advanced cash crop production. Hyden, op. cit. pp. 49, 56, 64-65 and passim.

[18] Srivastava and Livingstone, op. cit. Rich peasants began to arise after the second World War in Southern, Central and Eastern provinces in Zambia, but their position was and still remains weak by comparison with the large scale commercial, or rural capitalist sector. With the great bulk of the peasantry excluded then and now from governmental assistance, generalised agricultural backwardness has resulted; agricultural output is less than 14 percent of GDP. See, for example, Good, op. cit.

The settler colonies of Africa represent a quite dissimilar case. No articulation was ever initiated, but an advanced, settler owned commercial agriculture was developed on the basis of the extreme exploitation of land and labour and active support from a strong settler state. After independence small peasants in Kenya and Zimbabwe markedly increased their output of specific crops when assisted by the state to do so. Zimbabwe today has a capitalist agricultural system which combines very high productivities with continuing high inequalities in the ownership and distribution of productive resources, where agriculture is integrated with manufacturing industry within a broadly developed capitalist economy. These are special, colonial-historical features difficult, if not impossible, to replicate. In a certain sense, independent Zimbabwe and Kenya have yet to confront the deeper problems of the absence of a successful articulation of peasant and colonial agricultural systems.

[19] Hyden, op. cit., passim.

[20] Personal interviews, Goroka, 4 May 1977 and 6 July 1978.

[21] Near the end of the 1960s, Hari Gotoha was last on Finney's list of the ten leading Gorokan entrepreneurs. His personal assets then included 1,400 coffee trees, one trade store, two trucks and cattle, and Finney estimated the total value of Gotoha's property, part of which was held in partnership with Soso Subi, at $100,000. Finney, op. cit., p. 84 and 104-106. By the end of the 1970s, as noted, Gotoha was far richer.

[22] In 1976 115,887 farmers produced some 8,600 tonnes on a total of 11,760 hectares. Between 1973 and 1976 the total number of growers had increased by 114 per cent but the cultivated area had risen by only 47 per cent. Figures provided by J.J. Nitsche, personal interview, Department of Primary Industry, Goroka, 4 July 1977.

[23] The National Public Expenditure Plan 1978-1981, National Planning Office, Waigani, February 1978, p. 54.

[24] The average annual income of coffee growers in 1978 was said to be about K150. Post-Courier, 26 July 1979.

[25] Iliffe, op. cit., pp. 275, 289 and 293.

[26] Assefa Bequele, 'Stagnation and Inequality in Ghana', in Ghai and Radwan, op. cit., p. 226.

[27] Differentials with regard to land ownership, equipment and wage labour were also very large. Eddy Lee, 'Export-led Rural Development: the Ivory Coast', in Ghai and Radwan, op. cit.

[28] Helene Barnes, 'Women in Highlands Agricultural Production', in Denoon and Snowden, op. cit., pp. 275 and 280 where she refers to the studies of Anderson and others.

[29] Barnes, op. cit., p. 276 and chapter 6 above.

[30] Barnes, op. cit., p. 272.

[31] The average daily return in Goroka was some K2 - K3, and few women vendors in Goroka and Mt. Hagen came to market more than once a week. Ibid., p. 277.

[32] Quoted in ibid., p. 265.

[33] Anderson's survey of 1976, ibid., p. 275.

[34] Ibid., p. 277. Auwo Ketauwo's position is considered in chapters 5 and 6 above.

[35] Geoff Harris, editor of Yagl-Ambu, University of Papua New Guinea, 1978, and quoted in Barnes, op. cit., p. 279.

[36] A study by Allen in 1978 noted that women were 'working themselves into a state of near exhaustion seven days a week' in order to feed their families and pigs. Quoted in Barnes, op. cit. p. 280.